RESEARCH AND PERSPECTIVES IN ENDOCRINE INTERACTIONS

J.-C. Carel   P.A. Kelly   Y. Christen (Eds.)

Deciphering Growth

With 42 Figures and 14 Tables

 Springer

Jean-Claude Carel, M.D.
Hôpital Saint Vincent de Paul
INSERM U.561 –
Service d'Endocrinologie Pédiatrique
Groupe Hospitalier Cochin-Saint Vincent
de Paul-La-Roche Guyon
74 à 82, Av Denfert Rochereau
75674 Paris Cedex 14
France
E-mail: carel@paris5.inserm.fr

Yves Christen, Ph.D.
Fondation IPSEN
Pour la Recherche Thérapeutique
24, rue Erlanger
75781 Paris Cedex 16
France
E-mail: yves.christen@ipsen.com

Paul A. Kelly, Ph.D.
Inserm U584
Faculté de Médecine Necker
Enfants Malades
156, Rue de Vaugirard
75730 Paris Cedex 15
France
E-mail: kelly@necker.fr

ISSN 10   3-540-26192-3   Springer-Verlag Berlin Heidelberg New York
ISBN 13   978-3-540-26192-6   Springer-Verlag Berlin Heidelberg New York

Library of Congress Control Number: 2005926672

Springer-Verlag is a part of Springer Science+Business Media

springeronline.com

© Springer-Verlag Berlin Heidelberg 2005
Printed in Germany

Editor: Dr. Rolf Lange, Heidelberg
Desk Editor: Anne Clauss, Heidelberg
Production: PRO EDIT GmbH, 69126 Heidelberg, Germany
Cover design: design & production, 69121 Heidelberg, Germany
Typesetting: SDS, Leimen, Germany
Printed on acid-free paper   27/3152Re – 5 4 3 2 1 0

Preface

Growth is a complex process that is essential to life. Not only does size play an important role in the process of cellular proliferation, but body size is also a critical factor in determining which organisms live longer. Evolution has been characterised by a dramatic increase in an organism's body size, which is not only limited to the size of the brain.

In mammals, the major factors involved in the regulation of body growth are known: insulin-like growth factors (IGF) are key regulators of somatic growth. Growth hormone (GH), secreted by the pituitary gland, directly regulates circulating levels of IGF-I, which is the major coordinator of spatiotemporal growth of the organism. In humans, growth is even more complex, involving a number of specific characteristics not found in other species. These include rapid intrauterine growth, deceleration just after birth, a mid-childhood growth spurt, a second deceleration before puberty, an adolescent growth spurt, and finally full statural growth, which is seen somewhat later. The combined knowledge concerning the endocrine and paracrine aspects of growth have led to the introduction of treatment regimens, most effective in GH-deficient children. However, size depends on the combination of a number of genetic factors, and there remain several aspects of this complex process still poorly understood.

The Fondation Ipsen organized the fourth meeting in a series on *Research and Perspectives in Endocrine Interactions* entitled **Deciphering Growth** (Paris, December 6, 2004). The one-day meeting held in Paris was divided into three general sessions. The first, *Evolution and genetics of growth*, included talks on evolutionary trends in body size, the human evolution of gender-specific growth patterns, and the genetic control of body size at birth. The second general session was on the *GH – IGF-I axis*. Six speakers covered this broad subject, with presentations on isoforms of the GH receptor and growth in normal and pathological conditions, the use of mouse models to understand the role of the IGF-I receptor in growth, GH receptor downstream signalling, differential actions of GH and IGF-I in target tissues, the regulation of brain growth by IGF-I via direct effects on neural progenitor cell proliferation and survival, and phenotypes associated with human IGF-I gene deletion. The final session was on *Clinical approaches*. The first two speakers covered the clinical perspective of measurement of circulating IGF-I levels and the importance of mutations of the type 1 IGF receptor in IGF-I resistance. The last two talks described the importance of national growth registries to pediatric research, the French registry and the National Cooperative Growth Study (NCGS).

P. Kelly
J.C. Carel
Y. Christen

P.S. The editors wish to thank Mrs. Jacqueline Mervaillie for the organization of the meeting and Mrs. Mary Lynn Gage and Mrs. Astrid de Gérard for the editing of the book.

Contents

Contributors

Abuzzahab, Jennifer
Children's Hospitals and Clinics,
St. Paul, Minnesota,
USA

Bachelot, Anne
Inserm U584, Faculté de Médecine Necker,
156, Rue de Vaugirard, 75730 Paris Cedex 15,
France

Binart, Nadine
Inserm U584, Faculté de Médecine Necker,
156, Rue de Vaudirard, 75730 Paris Cedex 15,
France

Camacho-Hübner, Cecilia
Research Centre in Clinical and Molecular Endocrinology,
William Harvey Research Institute,
Queen Mary's School of Medicine and Dentistry,
Queen Mary's University of London, London,
UK

Carel, Jean-Claude
Pediatric Endocrinology and Inserm U561,
Groupe hospitalier Cochin-Saint Vincent de Paul,
82 av Denfert Rochereau, 75014 Paris,
France

Chaussain, Jean-Louis
Pediatric Endocrinology and Inserm U561,
Groupe hospitalier Cochin-Saint Vincent de Paul,
82 av Denfert Rochereau, 75014 Paris,
France

Chernausek, Steven D.
University of Cincinnati School of Medicine,
Cincinnati Children's Hospital Medical Center,
Division of Endocrinology,
3333 Burnet Ave., Cincinnati OH 45229,
USA

Clark, A.J.L.
Department of Endocrinology, Barts and the London,
Queen Mary's University of London, BEM Unit, London,
UK

Clayton, Peter E.
Academic of Child Health, Division of Human Development
and Reproductive Health, The Medical School, University of Manchester,
Manchester M13 9PT,
UK

Das, U.
Academic of Child Health, Division of Human Development
and Reproductive Health, The Medical School, University of Manchester,
Manchester M13 9PT,
UK

David, A.
Department of Paediatric Endocrinology, St Bartholomew's Hospital,
West Smithfield, London EC1A 7BE,
UK

De Magalhaes Filho, Carlos
Inserm U515, Hôpital Saint-Antoine, 75571 Paris 12,
France

Denley, Adam
School of Molecular and Biomedical Science, University of Adelaide,
Adelaide,
Australia

Dunger, David B.
Department of Paediatrics, University of Cambridge,
Addenbrooke's Hospital, Box 116 Level 8,
Hills Road, Cambridge CB2 2QQ,
UK

Frederick, Terra J.
Department of Neural and Behavioral Sciences, Penn State College of Medicine,
500 University Drive, Hershey PA 17033,
USA

Hintz, Raymond L.
Department of Pediatrics, Stanford University Medical Center, Rm S-302,
300 Pasteur Drive, Stanford CA 94305,
USA

Holzenberger, Martin
Inserm U515, Hôpital Saint-Antoine,
75571 Paris 12,
France

Hwa, Vivian
Department of Pediatrics, NRC5, Oregon Health and Sciences University,
3181 SW Sam Jackson Park Rd., Portland OR 97239-3098,
USA

Kappeler, Laurent
Inserm U515, Hôpital Saint-Antoine, 75571 Paris 12,
France

Kelly, Paul A.
Inserm U584, Faculté de Médecine Necker Enfants Malades,
156, Rue de Vaugirard, 75730 Paris Cedex 15,
France

Kiess, Wieland
Hospital for Children and ADolescents, Leipzig,
Germany

Le Bouc, Yves
Inserm U515, Hôpital Saint-Antoine, 75571 Paris 12,
France

Metherell, L.A.
Department of Endocrinology, Barts and the London,
Queen Mary's University of London, BEM Unit, London,
UK

Ness, Jennifer K.
Department of Neural and Behavioral Sciences, Penn State College of Medicine,
500 University Drive, Hershey PA 17033,
USA

Ong, K.K.
Department of Paediatrics, University of Cambridge,
Addenbrooke's Hospital, Box 116 Level 8, Hills Road, Cambridge CB2 2QQ,
UK

Orme, C. David L.
Division of Biology, Imperial College London, Silwood Park Campus,
Ascot SL5 7PY
UK

Osgood, Doreen
Hallet Center for Diabetes and Endocrinology, Brown Medical School,
Providence, Rhode Island,
USA

Pereira, L.A.
Department of Paediatric Endocrinology, St Bartholomew's Hospital,
West Smithfield, London EC1A 7BE,
UK

Petry, C.J.
Department of Paediatrics, University of Cambridge,
Addenbrooke's Hospital, Box 116 Level 8, Hills Road, Cambridge CB2 2QQ,
UK

Purvis, Andy
Division of Biology, Imperial College London, Silwood Park Campus,
Ascot SL5 7PY,
UK

Rosenfeld, Ron G.
Stanford University, Oregon Health and Science University,
Lucile Packard Foundation for Children's Health,
770 Welch Road, Suite 350, Palo Alto CA 94304,
USA

Savage, Martin O.
Department of Paediatric Endocrinology, St Bartholomew's Hospital,
West Smithfield, London EC1A 7BE,
UK

Schneider, Anke
Hospital for Children and ADolescents, Leipzig,
Germany

Smith, Robert J.
Hallet Center for Diabetes and Endocrinology, Brown Medical School,
Providence, Rhode Island,
USA

Sotiropoulos, Athanassia
Inserm U584, Faculté de Médecine Necker,
156, Rue de Vaugirard, 75730 Paris Cedex 15,
France

Whatmore, A.J.
Academic of Child Health, Division of Human Development
and Reproductive Health, The Medical School, University of Manchester,
Manchester M13 9PT,
UK

Walenkamp, Marie-José
Department of Pediatrics, Leiden University Medicalcenter,
P.O. Box 9600, 2300 RC Leiden,
The Netherlands

Wit, Jan-Maarten
Department of Pediatrics, Leiden University Medicalcenter,
P.O. Box 9600, 2300 RC Leiden,
The Netherlands

Wood, Teresa L.
Department of Neural and Behavioral Sciences, Penn State College of Medicine,
500 University Drive, Hershey PA 17033,
USA

Evolutionary Trends in Body Size

Andy Purvis[1] *and C. David L. Orme*[1]

Summary

An organism's body size tells us a lot about how it makes a living, suggesting that body size is a key parameter in evolution. We outline three large-scale trends in body size evolution. Bergmann's Rule is the tendency for warm-blooded species at high latitudes to be larger than their close relatives nearer the equator. The Island Rule is the trend for small species to become larger, and large species smaller, on islands. Cope's Rule, which we discuss in much more detail, is the tendency for lineages to increase in size over evolutionary time. Trends are best studied by combining data on evolutionary relationships among species with fossil information on how characters have changed through time. After highlighting some methodological pitfalls that can trap unwary researchers, we summarise evidence that Cope's Rule, while not being by any means universal, has operated in some very different animal groups – from microfauna (single-celled Foraminifera) to megafauna (dinosaurs) - and we discuss the possibility that natural selection and clade selection may pull body size in opposite directions. Despite size's central importance, there is little evidence that body size differences among related groups affect their evolutionary success: careful comparisons rarely reveal any correlation between size and present-day diversity. We end by touching on human impacts, which are often more severe on larger species.

Introduction

Body size matters. Whole books have been written on the many ways in which large organisms are very different from small ones (Peters 1983; Schmidt-Nielsen 1984; Brown and West 2000). Among many other differences, species of larger organisms are typically less abundant (Damuth 1993), live longer, and reach sexual maturity at a greater age (Millar and Zammuto 1983). Even within a single taxonomic group, such as mammals, the scale of life history differences between small and large species is immense. Table 1 contrasts bank voles (*Arvicola terrestris*) with African elephants (*Loxodonta africana*): a bank vole population could go through 17 generations in the time it takes to wean a newborn African elephant, and 51 by the time the newborn elephant itself becomes a parent (Pur-

---

[1] Division of Biology, Imperial College London, Silwood Park Campus, Ascot, SL5 7PY, U.K.

Carel et al.
Deciphering Growth
© Springer-Verlag Berlin Heidelberg

Table 1. Small mammals (such as bank voles) and large mammals (such as African elephants) differ in much more than size.

	Bank vole	African Elephant
Adult size (g)	17	276600
Weaning age (d)	20	1890
Age at first parturition (d)	58	5460
Litter size	4.9	1.06
Litters per year	3.4	0.14
Mortality, per month	0.27	0.005

vis and Harvey 1996). Going beyond anecdote, more inclusive comparative data sets show that many features of morphology, life history and ecology covary with adult body size. The relationships often provide a good fit to a power law, $y = aM^b$ (where M is mass and y the other variable of interest). Logarithmic transformation yields a straight line relationship, $\log y = \log a + b \cdot \log M$. The correlations of adult body mass with age at sexual maturity and population density across species of mammals are shown in Figure 1. Clearly, knowledge of a species' body size gives us very useful insights into how it makes its living. In this chapter, we view body size from an evolutionary perspective. Our concern is not with the *mechanisms* by which organisms achieve a particular size, but the *reasons*: we are concerned with ultimate rather than proximate causes.

Body size affects so many aspects of life that there are important costs and benefits to both large and small size. This is true both among individuals within species and at a higher level among species; Table 2 provides a far from exhaustive list. A species' mean body size is set through natural selection optimising the trade-off between costs and benefits. This process is perhaps clearest in groups like mammals that show deterministic growth (i.e., they more or less stop growing at sexual maturity); larger adults can invest more heavily in reproduction, but delaying the onset of reproduction to grow larger runs the risk of dying before maturity (Charnov 1991; Kozlowski and Weiner 1997). Under this view, adult mortality rates are important in determining adult body size – a point to which we shall return later. The body size distribution of a higher taxon will depend not only on the effects of natural selection within species but also on how body size affects the fates of those species. For instance, it has often been argued that species with short generation times are more able to evolve in response to environmental changes, and so are less likely to go extinct. Generation time correlates positively with body size, so such a process would cause small species to persist for longer and hence (other things being equal) to dominate numerically. Such macroevolutionary processes have long been controversial, but there is strong and growing evidence that selection does indeed operate among lineages (Williams 1992; Purvis 1996; Barraclough et al. 1998 b. Coyne and Orr 2004).

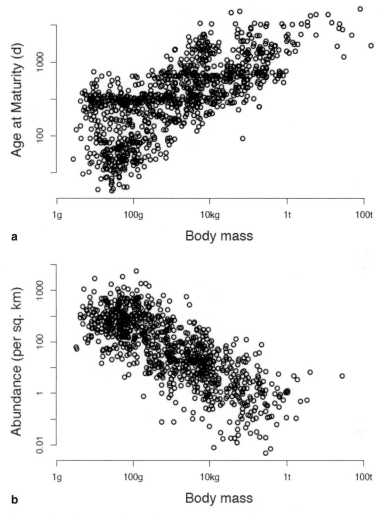

Fig. 1. Plots showing relationship across mammal species between adult body mass and (a) age at sexual maturity and (b) population density. Sample sizes are 1,030 species and 927 species, respectively, covering all mammalian orders. For both relationships, $r^2 > 0.5$. (Data from Jones et al., in preparation).

Given the ecological and evolutionary importance of body size, it is perhaps unsurprising that empirical patterns in body size have received so much attention. In this chapter, we first outline the three main general trends in body size that have been identified: Bergmann's Rule, the Island Rule, and Cope's Rule. Then we digress into different ways of testing for and understanding evolutionary trends, highlighting pitfalls that can trap the unwary. Focusing on Cope's Rule (the tendency for lineages to increase in body size through time), we then

Table 2. Possible advantages of large and small size to individuals and to species.

	Individual	Species
Big	Better competitor	Better competitor
	– in natural selection	
	– in sexual selection	
	Live longer	
	More buffered	
	Spend more on reproduction	
Small	Reproduce earlier	Rapid evolution
	Require less energy	Larger population
		More niches

show how a combination of fossil information can be combined with knowledge of the evolutionary relationships among species (phylogeny) to give insight into whether the trend was produced by within-lineage or among-lineage selection. We end with two further questions, one about today's biodiversity and the other about tomorrow's. Given the evidence of trends, is there a tendency for body size to correlate with diversity in the present-day biota? And how are human actions shaping body size trends?

Body size trends

Three more-or-less general body size trends have been claimed in evolution (though they have also had their detractors: Gould 1997; Gaston et al. 1998). Bergmann's Rule (Bergmann 1847) originally stated that warm-blooded verte-brate species from colder places tend to be larger than closely related species from warmer places; it has since been extended to cover within-species trends and trends within other taxa (see Ashton 2001 for a review). The broadest sur-veys and meta-analyses support the within-species version in mammals (espe-cially within larger species) and birds (Freckleton et al. 2003; Meiri and Dayan 2003).

Island systems have often been treated as natural experiments in evolution, including body size evolution. Island lineages often show very different body sizes from their mainland relatives. An elephant species endemic to Malta, *El-ephas falconeri*, stood only 1m tall at the shoulder: it was only 25% as tall (and about 1% as heavy) as its ancestral mainland species *E. antiquus* (Lister 1996). Flores man – thought to be an island endemic race of *Homo erectus* – was also about 1m tall (Brown et al. 2004). Dwarfing is not the only response to island liv-ing, however. Lomolino (1985) showed that, in mammals, the body size response depended upon initial size, with small species getting larger and large species getting smaller. For mammals of about 1 kg, there was no tendency for size to increase or decrease. Lineages of small birds also increase in size on islands, with lineages of large birds showing a (nonsignificant) trend to smaller size (Baillie 2001).

Bergmann's Rule and the Island Rule are concerned with timescales of a few years to perhaps a million years. Cope's Rule – the tendency for body size to increase along evolutionary lineages through geological time – can operate over much longer timescales. Cope's Rule was first postulated over a century ago (Cope 1896). It has been invoked in a wide range of taxa (Kingsolver and Pfennig 2004), though some, perhaps many, do not show the trend (Jablonski 1997). The causes of Cope's Rule, where it is found , are contentious: are size increases due to natural selection within a population favouring the largest individuals? Or are they due to species of larger individuals having higher speciation and/or lower extinction rates? Or do both microevolutionary and macroevolutionary processes contribute to the trend? Before we consider body size trends in three very different animal groups, we first make some general procedural points about how trends can most insightfully be analysed (see also Alroy 2000).

Identifying trends

The detection of evolutionary trends is not always straightforward. Two pitfalls in particular have commonly trapped researchers on the topic. A classic way to test for body size trends has been to compare the body size distributions of a lineage at two or more points through geological time. Differences in the central tendency (sometimes, any difference) are viewed as trends; increases are viewed as Cope's Rule. However, inspection of distributions alone gives no insight about the evolutionary process responsible for changes, or even whether the changes can usefully be termed trends (Alroy 2000). Figure 2a shows the body size distribution of a hypothetical clade at two points in its history. Two very different scenarios, with different evolutionary interpretations, could have produced this pattern. The first is Cope's Rule: within the clade, every lineage underwent a body size increase (Fig. 2b). The second scenario (Fig. 2c) has virtually no body size increase within lineages, but invokes differential survival and proliferation of three major clades: the smallest lineage has died out, the middle lineage has not changed, and a new large lineage has radiated. While the fates of these three lineages correlate with their body sizes, there is far too little replication (n=3 lineages) for any convincing demonstration that size (or anything else) determined fate. To discriminate between these two scenarios, the body size distributions are not enough. We also need to be able to trace lineages through time – we need knowledge of their phylogeny (evolutionary interrelationships – the "family tree" linking all life).

Many phylogenetic analyses of trends map present-day species characteristics onto the phylogeny linking those species and try thereby to test for the existence of trends. A common mistake is to infer a trend from a pattern like that seen in Figure 3a. Here, the most basal species (i.e., the one having fewest branching

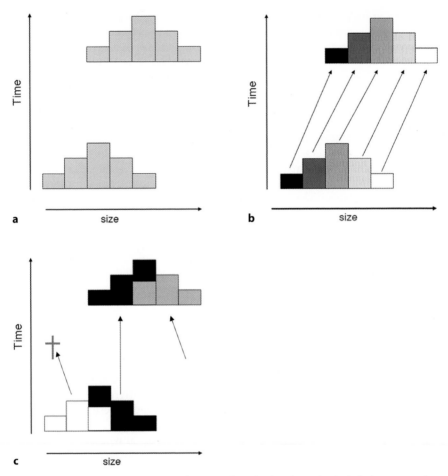

Fig. 2. a. Illustrative body size distributions of species in a hypothetical clade at two points in time. Differences in distribution are sometimes taken as evidence of evolutionary trends. **b.** Here, each lineage (represented by different shades of grey) has increased in body size: this is Cope's Rule. **c.** Here, the smallest lineage has gone extinct, the middle lineage survived unchanged, and a new large lineage has diversified. Distributions alone cannot distinguish between these two scenarios, so give little insight into the reality of, or process behind, evolutionary trends.

points between it and the most recent common ancestor of all the species) is the smallest. The inference of Cope's Rule, however, would be a logical error. Cope's Rule does not lead to any prediction about whether more basal extant species will be large or small, because all extant species have had equally long to evolve larger size (Fig. 3b). As well as a phylogeny, we need information about history – we need data for fossil species so we can trace body size along lineages through evolutionary time. [There is a range of methods that can be used to estimate ancestral body sizes (Swofford and Maddison 1987; Schluter et al. 1997; Garland

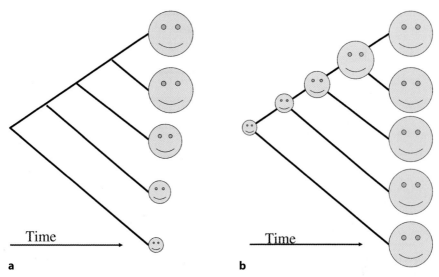

Fig. 3. Mapping body size of extant species onto a phylogeny may incorrectly identify trends. **a.** A phylogeny in which the basal species (having few nodes between them and the common ancestor of all the species) is smallest. Such a pattern is sometimes taken to support Cope's Rule. **b.** The pattern of body size to be expected if size has increased in all lineages: ancestors were small, but extant species are all large.

et al. 1999; Pagel 1999), which might seem to provide a way of looking for trends without using fossil data. However, such methods perform terribly in the presence of trends (Cunningham et al. 1998; Oakley and Cunningham 2000; Webster and Purvis 2002), totally undermining their use in tests for Cope's Rule.]

The ideal situation is to have a fossil record so densely sampled that every species can be traced from its start to its demise. Such a data set would let us test whether lineages tend to get larger or smaller over time and whether size affects probabilities of speciation and/or extinction. Very few groups have fossil records even approaching this level of quality, but one that does is the planktonic foraminifera (Pearson 1993), single-celled animals with both high abundance and good fossilisation potential. Webster and Purvis (2002) focused on a subset of species for which both a phylogeny (Fordham 1986) (Fig. 4a) and ancestral body sizes (of nodes labelled with letters on Fig. 4a) are known. This data set permitted 30 within-lineage comparisons of ancestral and descendant body sizes; significantly more than half of these showed size increases (21/30, sign test $p = 0.04$; Fig. 4b). This result provides strong evidence for Cope's Rule in this clade of protists (se also Arnold et al. 1995 for a broader analysis).

The fossil records of most groups are inadequate for such a detailed study but may still permit useful phylogenetic analyses. Alroy (1998) showed that body size has tended to increase through time within genera of North American mammals. He compared the size of newly appearing species with the size of putative ancestors (earlier species from within the same genus). In 442 of the

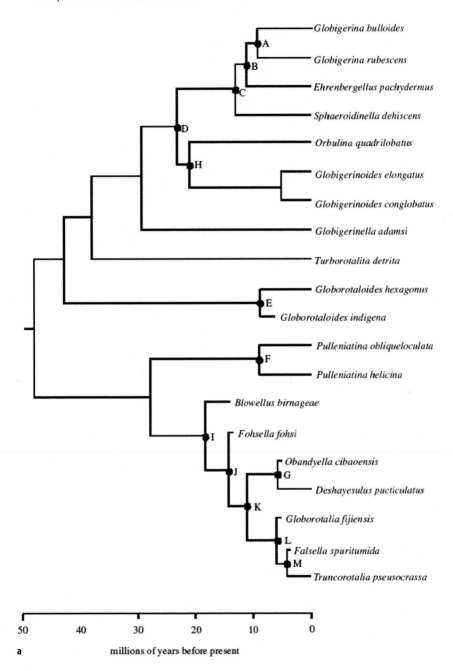

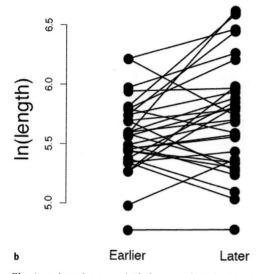

b   Earlier   Later

Fig. 4. a. (previous page) Phylogeny of 20 planktonic foraminiferan species. Letters indicate ancestral species whose body size was estimated directly from fossil specimens. After Webster and Purvis (2002). **b.** Results of 30 matched-pairs comparisons between ancestral and descendant species: 21 of the lineages show an increase in body size.

779 independent comparisons that were possible, the newly appearing species was larger than the putative ancestor (sign test: p = 0.0002), and the average body mass change was a 9.1% increase. Further analyses showed that the trend was consistent through the Cenozoic, but that the sudden size increase of some mammals immediately after the end-Cretaceous mass extinction was extreme (Alroy 1998). Alroy also compared the rate of within-genus size increase to the overall rate of size increase in mammals as a whole. The logic behind this comparison is that the within-genus comparisons provide a more direct window on the direction of evolution within lineages, whereas the overall rate of change reflects selection among lineages to a greater extent. If the two rate estimates differ significantly, it implies that both microevolution and macroevolution have contributed to the size trend (Alroy 1998; Hone et al. 2005). In mammals, the rate estimates were indistinguishable, making it possible that Cope's Rule results solely from microevolutionary processes.

Hone et al. (2005) used a similar method to investigate body size trends in dinosaur evolution. The earliest dinosaurs, such as *Coelophysis*, were typically small, whereas many of the famous large genera (e.g., *Tyrannosaurus*) date from the very end of the dinosaurs (more accurately, from the very end of non-avian dinosaurs – birds are descended from dinosaurs and so, in a sense, are dinosaurs themselves). Further, a regression of size on stratigraphic age across dinosaur genera shows a significant slope, with more recent genera being larger. This finding hints at a possible trend but, as shown above, is far from conclusive

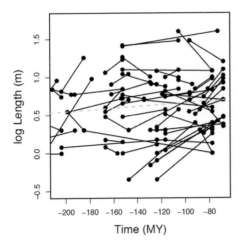

Fig. 5. Cope's Rule in dinosaurs. Points are genera (or groups of genera) of dinosaurs; solid lines link related lineages of different ages. The majority of these lines have positive slopes (indicating Cope's Rule). The average slope of the lines is significantly greater than the overall slope seen across all genera (grey dashed line). (MY: million years)

because it does not consider lineages. Combining data on the sizes and ages of dinosaur genera with a comprehensive phylogeny of the clade (Pisani et al. 2002) permitted construction of 65 independent comparisons between earlier and later related genera. Most of these comparisons showed size increasing through time (Fig. 5), in line with Cope's Rule. The size increase continued throughout dinosaur history, implying that Cope's Rule was operating in this group for some 165 million years. Intriguingly, the average slope of the within-lineage comparisons is significantly steeper than the regression line across all genera; individual lineages were growing faster than the dinosaur fauna as a whole. One explanation for such a discrepancy is that, while natural selection favoured larger individuals within species, clade selection (Williams 1992) favoured smaller species within clades (Alroy 2000). Smaller species may have speciated faster, or persisted for longer, than larger species. It is important to emphasise, however, that alternative explanations are also possible. In particular, underestimation of the true ages of genera could bias the within-lineage slopes more than the overall slope, leading to the observed difference. Quantitative data on fossil record quality would be needed to test this possibility (Hone et al. 2005).

A recent survey (Kingsolver and Pfennig 2004) sheds further light on the possible roles of within- and among-lineage selection in Cope's Rule. Of 854 collated estimates of the strength of directional selection on phenotypes of natural populations, the 91 selection gradients of traits relating directly to overall body size tended to be positive (i.e., selection favoured larger individuals), whereas the 763 other gradients showed no tendency to be positive or negative. Further, large size was advantageous for survival, fecundity and mating success, indicating that it was favoured by both natural and sexual selection (Kingsolver and Pfennig 2004). However, this within-lineage advantage of large size is not entirely consistent with the patterns seen in foraminiferans, mammals and dinosaurs. The directional selection for large size found by the survey is too strong: it would lead to very much more rapid size increase than ever seen in the fossil record

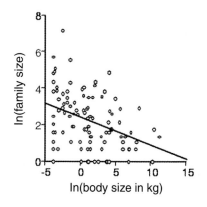

Fig. 6. Relationship between body size and diversity across mammalian families. A non-phylogenetic analysis (treating families as statistically independent) suggests a highly significant negative relationship. (After Purvis et al. 2003).

(Kingsolver and Pfennig 2004). Why the discrepancy? One possibility is that large size has disadvantages for the individual. For instance, large individuals tend to start reproducing later, increasing the chances of dying first (Millar and Zammuto 1983). Selection on age at maturity is indeed directional, pulling in the opposite direction to body size (Kingsolver and Pfennig 2004). The data are also compatible, however, with a scenario in which natural selection (favouring larger individuals) and clade selection (favouring smaller species) are acting in opposite directions (Kingsolver and Pfennig 2004).

Body size appears to matter to microevolution, and perhaps to macroevolution. Do large or small body size matter for evolutionary "success" of lineages? Good measures of lineages' evolutionary success are hard to come by, and present-day species diversity is a commonly used surrogate (Williams 1992; Coyne and Orr 2004). Does body size predict diversity?

Body size and diversity

At face value, looking for a correlation between body size and lineage diversity seems straightforward: simply collect body size data across a number of comparable lineages and plot the average body size within each lineage against the number of species in each lineage. This approach has been applied across a wide range of groups of organisms (e.g., Van Valen 1973) and typically reveals a strong negative trend: small-bodied lineages are, on average, more diverse. Figure 6 shows this relationship for mammalian families. Unfortunately, such an approach suffers from two potential problems, both of which result from failing to take the evolutionary relationships of species into account.

The first problem lies in the choice of *comparable* lineages. Studies have typically defined lineages of equal Linnean rank as their comparable units. Equal-ranked lineages are, however, not necessarily of equal age and phylogenies with a timescale have been used (Avise and Johns 1999) to demonstrate just how disparate their ages can be (or conversely, just how disparate the ranks of equal-aged lineages can be). This problem is exacerbated if evolutionary trends in body size

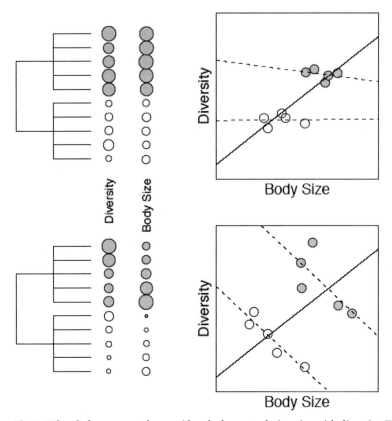

Fig. 7. Why phylogeny must be considered when correlating size with diversity. Two phylogenies are shown, each with the body size and diversity of the 10 groups indicated by the size of the symbols. The actual values are plotted on the right, with the solid line showing the size-diversity relationship when phylogeny is ignored and the dotted lines showing that relationship within the two major clades. In the upper phylogeny, the groups in one clade are both larger and more diverse, but there is no overall trend within the clades; in the lower phylogeny, there is an overall trend within both groups but it is in the opposite direction to that suggested by a non-phylogenetic approach. (After Purvis et al. 2003).

are occurring within lineages: given equal diversification rates between lineages, then lineages will accumulate diversity and changes in body size in proportion to their age.

The second problem arises because even equal-aged lineages are related to one another in an hierarchical way: each lineage does not represent an independent source of data because it shares varying amounts of evolutionary history with all other lineages in an analysis, as revealed by a phylogeny. If not accounted for, such non-independence can be seriously misleading (Nee et al. 1996; Barraclough et al. 1998a). Figure 7 shows hypothetical examples where non-phylogenetic analyses support a relationship between size and diversity where one does not exist or where the opposite relationship exists. This is not just a theoretical

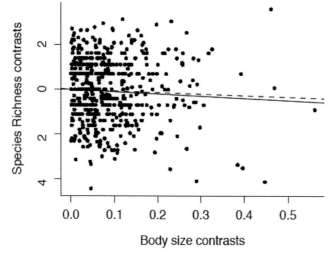

Fig. 8. Phylogenetic comparisons reveal no link between body size and diversity in either vertebrates (dashed line) or invertebrates (solid line).

problem; for example, the relationship between size and diversity in bird families (Van Valen 1973) has since been shown to result from phylogenetic non-independence (Nee et al. 1992).

Consequently, it is necessary for any test for a link between body size and diversity to use a phylogenetic framework. One approach (Felsenstein 1985; Mitter et al. 1988; Agapow and Isaac 2002) is to use a phylogeny of a group to identify sister clades: two lineages that are each other's closest relative. Such pairs of lineages are by definition of equal age. In addition, the evolutionary history of the two lineages makes for powerful comparisons; the lineages are likely to be broadly similar because they share a recent common ancestor; and the lineages have undergone independent evolution since that common ancestor. As a result, measures of differences in diversity and body size between sister clades are free of the problems of non-comparability and non-independence that affect non-phylogenetic methods. Using absolute diversity differences is problematic because the variance of the difference increases with the total diversity of the sister clades; it is more robust to use the log of the ratio of sister clade diversities (Isaac et al. 2003).

Sister clade comparisons have been used to explore the relationship between diversity and body size for a range of animal groups at a number of different taxonomic scales. Generally, the pattern of increased diversity at small size shown in non-phylogenetic studies is not supported by phylogenetic studies, most of which show no relationship (Nee et al. 1992; Gittleman and Purvis 1998; Owens et al. 1999; Katzourakis et al. 2001; Orme et al. 2002 a,b). The Carnivora are an exception; closer investigation reveals that this relationship may depend on clades within the group rather than on the order as a whole (Gittleman and Purvis 1998).

One attribute of the studies listed above is that they use diverse animal groups for which phylogenies resolved to a relatively low taxonomic level are available; hence large numbers of comparisons can be made. While large numbers of comparisons permit more powerful tests, diverse groups are also likely to be disparate and so other variables may confound any relationship between body size and diversity. Sister clade tests on small groups are less likely to be confounded, but individual tests lack power because they rely on fewer comparisons. An alternative to the studies of a single diverse group is therefore to use many less diverse groups and look for common trends in the sign or slope of the body size–diversity relationship. In addition, the compilation of new studies sheds light on whether there is a reporting bias in existing studies.

Orme et al. (2002 a) applied this approach to 38 complete species-level phylogenies from a wide range of animal groups; only one group showed a significant relationship between body size and diversity, and neither the sign nor the slope of the relationship across the 38 groups was significantly different from zero. A possible criticism of this result is that the 38 groups are themselves related hierarchically; there may be a phylogenetic pattern within the set of slopes but, if the relationships between the groups are known, then this pitfall can be addressed. Orme et al. (2002a), using a phylogeny of the 38 groups, identified each of the nested sets of groups within the data set. Each nested set then allowed two questions to be asked about that clade: 1) do the pooled comparisons for the member groups show an overall size–diversity relationship, and 2) do the individual slopes for member groups differ significantly. Figure 8 shows the results for the most nested set of groups, which compares the vertebrate and invertebrate groups in the study. There is neither any significant overall relationship using the full set of pooled comparisons nor any significant difference between the slopes of the size–diversity relationship for vertebrates and invertebrates.

Taken as a whole, there therefore seems to be very little evidence to support a general link between small body size and elevated rates of diversification, although a few individual clades do show such a link. The evolutionary and ecological implications of these findings are discussed in detail by Purvis et al. (2003).

Human impacts

Although it is by no means universal (Jablonski 1997), Cope's Rule held sway in dinosaurs over 165 million years. Since dinosaurs died out, the rule has been evident in mammals. However, human impacts on ecosystems and species are cumulatively so severe that we may be reversing the trend: we are "downsizing nature" (Lomolino et al. 2001).

Hunting and harvesting increase individual mortality rates. As noted earlier, adult mortality rates are thought to determine adult body size for many species. In mammals, for instance, natural selection optimises the trade-off between surviving to reproduce (easier if you mature at an early age and hence are small) and investing in offspring production once mature (easier to do if you mature at a large size). Increasing mortality changes the optimal strategy: waiting to grow larger becomes increasingly risky, and selection favours earlier matura-tion. There is evidence that human hunting has indeed reduced age (and size) at maturity in both mammals (Law 2001) and fish (Reynolds et al. 2001).

Mammals domesticated for food (but not those domesticated for power) have also tended to dwarf (Purvis 2001). The smallest domesticated species actually show a size increase, but this is probably explained by the fact that such species typically have very high natural mortality rates: domestication as food animals may have increased their life expectancy by reducing predation.

Dwarfing is an adaptive response to human-caused mortality, but not all spe-cies have been able to adapt. Humans have also been agents of selection among lineages: we have caused extinctions of many large species, whose smaller rela-tives are still extant (e.g., Walker 1967). Among extant species of birds and mam-mals, larger species are more likely to be in rapid decline than smaller species (Bennett and Owens 1997; Purvis et al. 2000). In both cases, body size per se may not be the target of selection, but large species are more likely to have the long generation times, low reproductive rates, low abundances and so on that make them vulnerable.

Conclusions

There are no hard-and-fast empirical rules in evolution, because evolving popu-lations have to respond to a complex and ever-changing world. The complex-ity and fluidity of natural systems make it perhaps all the more surprising that there are any "rules of thumb," trends general enough to be given names, despite their many exceptions. Yet Bergmann's Rule, the Island Rule and Cope's Rule are useful rules of thumb, at least in certain groups and at certain times.

This chapter has shown how the detection and analysis of evolutionary trends in body size is best achieved by combining information from fossils with infor-mation on the evolutionary relationships among lineages. The historical analysis of lineages provides the most powerful framework for teasing apart the contribu-tions to the trend of natural selection within lineages and clade selection among lineages (Alroy 2000). There is increasing (but imperfect) evidence that selection may operate in different directions at these different hierarchical levels.

Despite the central importance of body size for species' ecology, life history and morphology, body size is a very poor predictor of diversification rates. It is possible that small size is a necessary but not sufficient condition for extremely rapid diversification (it is hard to imagine a world in which elephants are as di-verse as insects), but the lack of any tighter relationship is striking. It also places the current anthropogenic extermination of many large species into shocking perspective. At current rates of "progress," it may take humanity only a few cen-turies to undo the changes in mammalian body size distribution that took Na-ture tens of millions of years to achieve.

Acknowledgements

We thank Fondation Ipsen and the organisers for inviting us to participate in this book and the conference, and the Natural Environment Research Council (U.K.) for support through research grants NER/A/S/2001/00581 (A.P.) and NER/O/S/2001/01230 (C.D.L.O.).

References

Agapow P-M, Isaac NJB (2002) MacroCAIC: revealing correlates of species richness by comparative analysis. Div Dist 8: 41-43

Alroy J (1998) Cope's Rule and the dynamics of body mass evolution in North American fossil mammals. Science 280: 731-734

Alroy J (2000) Understanding the dynamics of trends within evolving lineages. Paleobiology 26: 319-329

Arnold AJ, Kelly DC, Parker WC (1995) Causality and Cope's Rule: evidence from the planktonic Foraminifera. J Paleontol 69: 203-210

Ashton KG (2001) Are ecological and evolutionary rules being dismissed prematurely? Div Dist 7: 289-295

Avise JC, Johns GC (1999) Proposal for a standardized temporal scheme of biological classification for extant species. Proc Natl Acad Sci USA 96: 7358-7363

Baillie JEM (2001) Persistence and vulnerability of island endemic birds. Ph.D. thesis, University of London

Barraclough TG, Nee S, Harvey PH (1998a) Sister-group analysis in identifying correlates of diversification - Comment. Evol Ecol 12: 751-754

Barraclough TG, Vogler AP, Harvey PH (1998b) Revealing the factors that promote speciation. Phil Trans R Soc Lond B 353: 241-249

Bennett PM, Owens IPF (1997) Variation in extinction risk among birds: chance or evolutionary predisposition? Proc R Soc Lond B 264: 401-408

Bergmann C (1847) Ueber die verhältnisse der wärmeökonomie der thiere zu ihrer grösse. Gottinger studien 3: 595-708

Brown JH, West GB, Eds. (2000) Scaling in biology. Oxford, Oxford University Press

Brown P, Sutikna T, Morwood MJ, Soejono RP, Jatmiko, Saptomo EW, Due RA (2004) A new small-bodied hominin from the Late Pleistocene of Flores, Indonesia. Nature 431: 1043-1044

Charnov EL (1991) Evolution of life history variation among female mammals. Proc Natl Acad Sci USA 88: 1134-1137

Cope ED (1896) The primary factors in organic evolution. Open Court Publishing Co., Chicago

Coyne JA, Orr HA (2004) Speciation. Sinauer, Sunderland MA

Cunningham CW, Omland KW, Oakley TH (1998) Reconstructing ancestral character states: a critical reappraisal. Trends Ecol Evol 13: 361-366

Damuth J (1993) Cope's rule, the island rule, and the scaling of mammalian population density. Nature 365: 748-750

Felsenstein J (1985) Phylogenies and the comparative method. Am Nat 125: 1-15

Fordham BG (1986) Miocene-Pleistocene planktic Foraminifers from DSDP sites 208 and 77, and phylogeny and classification of Cenozoic species. Evolutionary Monographs 6: 1-200

Freckleton RP, Harvey PH, Pagel M (2003) Bergmann's Rule and body size in mammals. Am Nat 161: 821-825

Garland T, Midford PE, Ives AR (1999) An introduction to phylogenetically based statistical methods, with a new method for confidence intervals on ancestral values. Am Zoolog 39: 374-388

Gaston KJ, Blackburn TM, Spicer JI (1998) Rapoport's rule: time for an epitaph? Trends Ecol Evol 13: 70-74

Gittleman JL, Purvis A (1998) Body size and species richness in primates and carnivores. Proc R Soc Lond B 265: 113-119

Gould SJ (1997) Cope's rule as psychological artefact. Nature 385: 199-200

Hone DWE, Keesey TM, Pisani D, Purvis A (2005) Macroevolutionary trends in the Dinosauria: Cope's Rule. J Evol Biol 18: 587-595

Isaac NJB, Agapow P-M, Harvey PH, Purvis A (2003) Phylogenetically nested comparisons for testing correlates of species-richness: a simulation study of continuous variables. Evolution 57: 18-26

Jablonski D (1997) Body-size evolution in Cretaceous molluscs and the status of Cope's rule. Nature 385: 250-252

Katzourakis A, Purvis A, Azmeh S, Rotheray G, Gilbert F (2001) Macroevolution of hoverflies (Diptera: Syrphidae): the effect of using higher-level taxa in studies of biodiversity, and correlates of species richness. J Evol Biol 14: 219-227

Kingsolver JG, Pfennig DW (2004) Individual-level selection as a cause of Cope's Rule of phyletic size increase. Evolution 58: 1608-1612

Kozlowski J, Weiner J (1997) Interspecific allometries are byproducts of body size optimization. Am Nat 149

Law R (2001) Phenotypic and genetic changes due to selective exploitation. In: Reynolds JD, Mace GM, Redford KH, Robinson JG (eds) Conservation of exploited species. Cambridge University Press, Cambridge, pp 323-342

Lister AM (1996) Dwarfing in island elephants and deer: processes in relation to time of isolation. Symp Zool Soc Lond 69: 277-292

Lomolino MV (1985) Body sizes of mammals on islands: the island rule re-examined. Am Nat 125: 310-316

Lomolino MV, Channell R, Perault DR, Smith GA (2001) Downsizing Nature: Anthropogenic dwarfing of species and ecosystems. In: Lockwood JL, McKinney ML (eds) Biotic homogenization. Kluwer Academic/Plenum Publishers, New York, pp 223-243

Meiri S, Dayan T (2003) On the validity of Bergmann's rule. J Biogeog 30: 331-351

Millar JS, Zammuto RM (1983) Life histories of mammals: an analysis of life tables. Ecology 64: 631-635

Mitter C, Farrell B, Wiegmann B (1988) The phylogenetic study of adaptive zones: has phytophagy promoted insect diversification? Am Nat 132: 107-128

Nee S, Mooers AØ, Harvey PH (1992) The tempo and mode of evolution revealed from molecular phylogenies. Proc Natl Acad Sci, USA 89: 8322-8326

Nee S, Barraclough TG, Harvey PH (1996) Temporal changes in biodiversity: detecting patterns and identifying causes. In: Gaston KJ (eds) Biodiversity: a biology of numbers and difference. Blackwell, Oxford, pp 230-252

Oakley TH, Cunningham CW (2000) Independent contrasts succeed where ancestor reconstruction fails in a known bacteriophage phylogeny. Evolution 54: 397-405

Orme CDL, Isaac NJB, Purvis A (2002a) Are most species small? Not within species-level phylogenies. Proc R Soc Lond B 269: 1279-1287

Orme CDL, Quicke DLJ, Cook J, Purvis A (2002b) Body size does not predict species richness among the metazoan phyla. J Evol Biol 15: 235-247

Owens IPF, Bennett PM, Harvey PH (1999) Species richness among birds: body size, life history, sexual selection or ecology? Proc R Soc Lond B 266: 933-939

Pagel M (1999) Inferring the historical patterns of biological evolution. Nature 401: 877-884

Pearson PN (1993) A lineage phylogeny for the Paleogene planktonic foraminifera. Micropaleontology 39: 193-232

Peters RH (1983) The ecological implications of body size. Cambridge University Press, Cambridge

Pisani D, Yates A, Langer MC, Benton MJ (2002) A genus-level supertree of the Dinosauria. Proc R Soc Lond B 269: 915-921

Purvis A (1996) Using interspecific phylogenies to test macroevolutionary hypotheses. In: Harvey PH, Leigh Brown AJ, Maynard Smith J, Nee S (eds) New uses for new phylogenies. Oxford Univ. Press, Oxford, pp 153-168

Purvis A (2001) Mammalian life histories and responses of populations to exploitation. In: Reynolds JD, Mace GM, Redford KH, Robinson JG (eds) Exploited species. Cambridge University Press, Cambridge, pp 169-181

Purvis A, Harvey PH (1996) Miniature mammals: life-history strategies and evolution. Symp Zool Soc Lond 69: 159-174

Purvis A, Gittleman JL, Cowlishaw G, Mace GM (2000) Predicting extinction risk in declining species. Proc R Soc Lond B 267: 1947-1952

Purvis A, Orme CDL, Dolphin K (2003) Why are most species small-bodied? A phylogenetic view. In: Blackburn TM, Gaston KJ (eds) Macroecology: concepts and consequences. Blackwell Science, Oxford, pp 155-173

Reynolds JD, Jennings S, Dulvy NK (2001) Life histories of fishes and population responses to exploitation. In: Reynolds JD, Mace GM, Redford KH, Robinson JG (eds) Conservation of exploited species. Cambridge University Press, Cambridge, pp 147-168

Schluter D, Price T, Mooers AØ, Ludwig D (1997) Likelihood of ancestor states in adaptive radiation. Evolution 51: 1699-1711

Schmidt-Nielsen K (1984) Scaling: why is animal size so important? Cambridge University Press, Cambirdge

Swofford DL, Maddison WP (1987) Reconstructing ancestral character states under Wagner parsimony. Math Biosci 87: 199-229

Van Valen L (1973) Body size and numbers of plants and animals. Evolution 27: 27-35

Walker AC (1967) Patterns of extinctions among the subfossil Madagascan lemuroids. In: Martin PS, Wright HEJ (eds) Pleistocene extinctions. Yale University Press, New Haven, pp 425-432

Webster AJ, Purvis A (2002) Testing the accuracy of methods for reconstructing ancestral states of continuous characters. Proc R Soc Lond B 269: 143-149

Williams GC (1992) Natural selection: domains, levels, challenges. Oxford University Press, Oxford

Sexual Dimorphism in the Growth of *Homo sapiens*: Facts, Inferences and Speculation

Ron G. Rosenfeld[1]

Summary

Sexual dimorphism in body size is a common feature of most animal species. While in many species, the female is the larger sex, in mammals, males are commonly larger, and greater male body mass and height are typical of primates. Growth of *H. sapiens*, however, is characterized by a number of unusual features, including rapid *in utero* growth, a prolonged childhood phase, a pubertal spurt in stature, and a relative lack of sexual dimorphism, with adult male height averaging only 107% of that of females. The recent report of a female patient with a homozygous mutation of the gene for STAT5b, a critical component of the growth hormone (GH) signaling cascade responsible for insulin-like growth factor (IGF) gene transcription, has demonstrated that growth in both females and males is strongly pulsatile GH-STAT5b-IGF-dependent. This common dependence of both human females and males on the GH-STAT5b-IGF pathway may explain the relative lack of dimorphic growth characteristic of *H. sapiens*.

Introduction

The size of organisms currently inhabiting the earth ranges over 21 orders of magnitude, from mycoplasma, with an average size of $<10^{-13}$ grams, to the blue whale, with a mass $>10^8$ grams. This extraordinary diversity of size demonstrates the power of evolutionary forces to generate what Darwin, himself, in the closing lines of The Origin of Species, identified as "endless forms most beautiful and most wonderful" (Darwin 1859).

Size variation is observed not only among species but also between sexes of a particular species. The extent of sexual dimorphism in any species reflects the difference between the sum of all selective pressures affecting the size of the female and the sum of all those affecting the male, as environmental pressures may impact size in each sex discordantly or in parallel. The most common state in the animal kingdom is for females to be larger than males, reflecting the energy needs necessitated by egg production and maintenance. In birds and

[1] Lucile Packard Foundation for Children's Health, Stanford University, Oregon Health and Science University

Carel et al.
Deciphering Growth
© Springer-Verlag Berlin Heidelberg

mammals, however, males are commonly the larger sex. A smaller female size represents energy reallocation from body mass to litter mass, a critical investment for the female to make, as reflected in the observation that a positive correlation remains between maternal body mass and mean mass of both individual progeny and the entire litter.

Although most studies of sexual dimorphism have focused on accentuated growth in the male, discrepancies in male:female ratios can result from differential growth of either sex. While selective pressures can affect both sexes of a species, differences in the roles of males and females within a species may preferentially impact one sex relative to the other. These factors include, but are not limited to, energy allocation for reproduction vs. survival, environmental factors (for primates, for example, arboreal vs. terrestrial life-styles), defense against predation, and intraspecies competition (for both limited resources and sexual mates). Darwin emphasized the latter aspect, pointing out that species with high male intrasexual competitition and high degrees of polygyny tended to exhibit more size dimorphism (Darwin 1871; Short 1994). Thus, in polygynous mammals, such as sea lions and gorillas, adult males may be greater than twice as large as females. Dimorphism is thereby predicated upon several assumptions that appear to hold firm for such species: 1) variation in reproductive success of males may be pronounced in polygynous species; 2) such variation is associated with increased competition among males for reproductive partners; and 3) increased competition selects for characteristics in males that serve to increase their likelihood of success in competition for mates. An additional implication of these assumptions is that females of a species tend to have the more "optimal" size for survival and that selection of differential growth in males is driven by enhanced reproductive success.

Assessment of dimorphism among species is further complicated by the observation that the degree of dimorphism may correlate with the size of the species. Rensch's Law of Sexual Dimorphism states that, as a general rule, larger species tend to show a greater degree of sexual dimorphism than do smaller species (Martin et al. 1994). Other factors that contribute potentially to dimorphic growth include not only different rates of growth between the sexes but also differential durations of growth phases in males vs. females, a factor that is clearly relevant to dimorphic growth in *H. sapiens* (see below).

Sexual dimorphism in primates

Although many primate species are characterized by pronounced sexual dimorphism, this is not universal. Little dimorphism is found in prosimian primates (i.e., lemurs, lorises and tarsiers) and New World monkeys; sexual dimorphism in size is largely restricted to Old World monkeys and apes (Martin et al. 1994; Fig. 1). The mandrill is the most dimorphic of the primates, with adult males frequently weighing up to three times more than adult females. Among the great apes, the adult male gorilla has a body weight 240% of the counterpart female, the male orangutan weight is 210% of the counterpart female, and the male chimpanzee weight averages 130% of the counterpart female. Although differences in

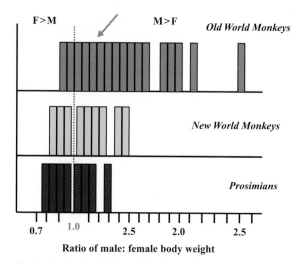

Fig. 1. Ratio of male:female body weight in three different groups of primates, demonstrating that marked sexual dimorphism is largely restricted to Old World monkeys and apes. The dotted line is drawn through a male:female ratio of 1.0. The arrow indicates the mild sexual dimorphism characteristic of *H. sapiens*. (Derived from Martin et al. 1994).

stature are less well documented in primate species, it is clear that dimorphism extends to height, as well. Dimorphism in primates may have been facilitated by the transition from a largely arboreal to a primarily terrestrial existence, as the latter supports a larger body size. Furthermore, the loss of access to tree-climbing might have selected for larger body size as a defense against predators.

Among the apes, size dimorphism appears to correlate with social structure and, interestingly, inversely with testicular volume (Short 1994; Fig. 2; Table 1). It has been inferred that among apes,- where a dominant male controls reproductive access to a harem primarily on the basis of his larger size - testicular volume and, by implication, spermatogenic capacity become less important variables. On the other hand, in multi-male mating societies, dominant size is of lesser significance in access to females, especially when several males may mate with a single female when she is in oestrus. In this situation, so-called "sperm wars" favor the male with the highest number of functional spermatozoa. It is of interest to ask where *H. sapiens* fits into this picture. As discussed below, humans are relatively nondimorphic compared to other apes, have medium testicular volume, and are not inherently monogamous. Roger Short has concluded that, "the best guess would be that we are basically a polygynous primate in which the polygyny usually takes the form of serial monogamy" (Short 1994).

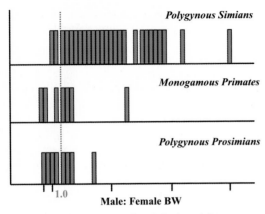

Fig. 2. Ratio of male:female body weight in three different primate social structures. Marked sexual dimorphism is largely restricted to polygynous simians. (Derived from Martin et al. 1994).

Table 1. Sexually dimorphic characteristics of primates, relative to social structure.

Social Structure	Testicular Volume	Size Dimorphism	Canine Dimorphism
Monogamy	+	+/-	+/-
Competing Males	+	++++	++++
Polyandrous Group-living	+++	++	++

Sexual dimorphism in *H. sapiens*

Although it is useful to relate the auxology and dimorphism of *H. sapiens* to that of other primates, it soon becomes evident that our species has a number of distinguishing features. The human growth curve, for starters, is extraordinarily complex, and it is not unreasonable to inquire what selective advantages these unique features might have conferred on the species: 1) the attainment of maximal growth velocity during late gestation, rather than in early infancy, as in most species; 2) deceleration of growth velocity after birth; 3) relatively late sexual maturation; 4) onset of puberty at the time of slowest growth in childhood; 5) the presence of a marked adolescent growth spurt in height; and 6) relatively prolonged delay between puberty and attainment of full reproductive capacity (particularly in females; Bogin 1999.

When compared with other primates, *H. sapiens* is characterized by a relative lack of sexual dimorphism. This is particularly true when Rensch's Law is taken into consideration, as *H. sapiens* is a relatively large primate and Rensch's Law would predict that the species would be characterized by more pronounced dimorphism. The mean difference in adult stature between males and females is only 12.6 cm (5 inches), reflecting only 7% of mean adult stature. Furthermore, prepubertally, males and females grow at essentially identical rates, with the difference in height between prepubertal boys and girls averaging only 1 cm. Essentially all sexual dimorphism in stature between male and female humans can be explained by differential growth during puberty, particularly in the timing of epiphyseal fusion under the influence of estrogen exposure. Thus, Longo et al. (1978) have accounted for this 12.6 cm difference in adult male vs. female stature in the following manner: 1) greater male growth prior to adolescence (+1.6 cm); 2) delayed onset of adolescence in males (+6.4 cm); 3) greater intensity of the male pubertal growth spurt (+6.0 cm); and 4) longer duration of growth after the pubertal growth spurt in females (-1.4 cm).

These two characteristics, the relative lack of sexual dimorphism and the presence of a pubertal growth spurt, are important distinguishing characteristics of human auxology and raise interesting questions concerning the selective advantages of these growth patterns, as well as the molecular mechanisms underlying them. The presence of a pronounced pubertal growth spurt in *H. sapiens* is unique among mammals, even when compared with other primates, who may have pubertal weight gains but little, if any, pubertal acceleration of statural growth. It is of value, therefore, to explore the potential biochemical and hormonal bases for dimorphic growth and inquire what mechanisms may have led to the mitigation of dimorphism in *H. sapiens*.

Molecular and biochemical basis for dimorphic growth

Given the critical role of the GH-IGF axis in postnatal mammalian growth, a search for a molecular explanation for dimorphic growth should begin there. Differences in growth between the sexes of a species may reflect differences at any of the following levels:
1) GH secretion
2) GH receptor concentrations
3) post-GH receptor signaling
4) IGF production
5) IGF responsiveness
6) factors outside the GH-IGF axis
7) timing of epiphyseal fusion

In the rat, a distinctly dimorphic species, GH secretion in males is pulsatile, with sharp peaks and deep troughs and a periodicity of every three to four hours (Tannenbaum and Martin 1976). Female rats are characterized by "continuous secretion" of GH; peaks and troughs occur but are less pronounced than in the male (Robinson et al. 1998). These differences in GH secretory patterns correspond to sexually dimorphic responses to GH that cannot be attributed solely to sex-based differences in the cumulative amount of GH secreted (Jansson et al. 1985; Waxman et al. 1991). Sexually dimorphic responses to GH in rats include differences in body growth, the induction of major urinary proteins in male rats by pulsatile GH secretion, and the induction of hepatic prolactin and GH receptors in female rats by continuous GH secretion. In prepubertal humans, however, no differences in GH secretory patterns have been identified, with pulsatile secretion observed in both sexes. Additionally, there is no evidence that GH receptor concentrations differ in prepubertal males and females. These observations are consistent with the near identical growth of boys and girls prior to the adolescent growth spurt.

GH stimulation of IGF-I gene transcription is mediated primarily through the Janus kinase-signal transducer and activator of transcription (Jak-STAT) pathway. Seven human STAT proteins have been identified but, as described below, evidence supports the conclusion that STAT5b is the critical mediator of IGF-I gene transcription by GH. Interestingly, targeted disruption of STAT5b in mice results in no alteration in the growth of female mice (Udy et al. 1997). Male mice with STAT5b knockouts, however, show a reduction in body size to that of wild-type females, together with a decrease in serum IGF-I concentrations to female levels and a shift of male-specific liver gene expression to female levels. These observations lead to the inference that, at least in rodents, STAT5b is the major determinant of sexual dimorphism in both body growth and hepatic gene expression, and that STAT5b, in turn, is regulated by the pulsatile vs. continuous GH secretory patterns of normal rodents. Indeed, Tannenbaum et al. (2001) have demonstrated that spontaneous GH secretory episodes elicit corresponding changes in STAT5 DNA-binding activity. Pulsatility of STAT activation, regulated by episodic GH secretion, could lead to gender-specific IGF production and growth patterns.

The relevance of these observations to human growth is, however, uncertain. As discussed above, prepubertal growth of boys and girls is remarkably similar, and the 7% difference in adult height is attributable to differential growth during puberty. Prepubertal boys and girls both have pulsatile GH secretion and, unlike the situation in rodents, STAT5b appears to be critically involved in the growth of both sexes. The latter point is underscored dramatically by the recent report of the first human mutation of the STAT5b gene, which occurred in a female patient and resulted in severe growth failure and profoundly low serum IGF-I concentrations, despite elevated GH levels (Kofoed et al. 2003). A second, as yet unreported case of a homozygous STAT5b mutation has been observed in a female patient, with similar severe growth retardation and reduction of serum IGF-I concentrations. These cases stand in marked contrast to the studies performed in rodents, where targeted deletion of the STAT5b gene resulted in impaired growth only in males.

These unique cases of human STAT5b deficiency have important implications for our understanding of human growth and dimorphism (Rosenfeld and Nicodemus 2003; Rosenfeld 2004). Since the degree of growth failure mirrors that observed in cases of GH insensitivity resulting from mutations of the GH receptor, it seems reasonable to conclude that essentially all of the growth-promoting actions of GH in humans are mediated through STAT5b regulation of IGF-I gene transcription (Rosenfeld et al. 1994). It is also apparent that normal (or even increased) levels of STAT5a cannot compensate for the deficiency of STAT5b, at least in terms of IGF-I gene transcription and skeletal growth (Teglund et al.

1998). Finally, given that the first two cases of homozygous mutations of the STAT5b gene occurred in females with profound growth failure, it is apparent that, in humans, **both** male and female growth is STAT5b-dependent. Indeed, the data support the hypothesis that human growth in both sexes is profoundly pulsatile GH-STAT5b-IGF-dependent.

In this light, it is of interest to look at the unique characteristics of human growth enumerated above and to speculate on the selective advantages that may have been garnered from these patterns. The shift from a primarily arboreal to a terrestrial existence for our hominid ancestors made increasing body size not only feasible but, perhaps, preferable. The rapid growth of the human fetus in utero may reflect, in part, its relatively premature birth, necessitated by the increase in cranial volume characteristic of *H. sapiens*. Indeed, from the time of *A. afarensis* until *H. sapiens* (a span of approximately 3.5 million years), the average endocranial volume increased from 438 to 1350 cm³. The prolonged childhood of *H. sapiens* is consistent with the need for a lengthy nurturing and learning phase, during which it might be advantageous for the child to have decreased food requirements and be perceived as non-threatening by adults of the species. The pubertal growth spurt could then be viewed as necessary to allow for rapid attainment of the ideal adult stature required for defense against predation, effective hunting and sexual selection, following the prolonged childhood phase.

Finally, the development in *H. sapiens* of a social structure less dependent upon a dominant male and more consistent with multimale:multifemale or monogamous relationships may have undermined the selective advantage of large body mass in the male of the species. At the same time, growth in the female may have been advantageous for her own survival, as well as the ability to accommodate and deliver a relatively large fetus with a dramatically increased endocranial volume.

While much of this discussion must be viewed as highly speculative, it remains an important exercise to view growth characteristics of any species in terms of the selective advantages that accrue to both individuals and the species as a whole. The molecular basis for growth and maturation of *H. sapiens* is likely to have evolved in a manner that maximizes the reproductive success and survival of the species.

References

Bogin B (1999) Patterns of human growth. 2nd Ed. Cambridge University Press, Cambridge

Darwin C (1859) The Origin of Species. 1st Ed. John Murray, London

Darwin C (1871) The Descent of Man, and Selection in Relation to Man. 1st Ed. John Murray, London

Jansson JO, Eden S, Isaksson O (1985) Sexual dimorphism in the control of growth hormone secretion. Endocr Rev 6:128-150

Kofoed EM, Hwa V, Little B, Woods KA, Buckway CK, Tsubaki J, Pratt KL, Bezrodnik L, Jasper H, Tepper A, Heinrich JJ, Rosenfeld RG (2003) Growth hormone insensitivity associated with a STAT5b mutation. New Engl J Med 349: 1139-147

Longo RG, Gasser TH, Prader A, Stutzle W, Huber PJ (1978) Analysis of the adolescent growth spurt using smoothing spline functions. Ann Human Biol 5: 421-434

Martin RD, Willner LA, Dettling A (1994) The evolution of sexual size dimorphism in primates. In: Short RV, Balaban E (eds) The differences between the sexes. Cambridge University Press, Cambridge, pp 159-200

Robinson IC, Gevers EF, Bennett PA (1998) Sex differences in growth hormone secretion and action in the rat. GH IGF Res 8: 39-47

Rosenfeld RG (2004) Gender differences in height: an evolutionary perspective. J Pediatr Endocrinol Metab 17(Suppl 4) 1267-1271

Rosenfeld RG, Nicodemus BC (2003) The transition from adolescence to adult life: physiology of the "transition" phase and its evolutionary basis. Horm Res 60(Suppl 1): 74-77

Rosenfeld RG, Rosenbloom AL, Guevara-Aguirre J (1994) Growth hormone (GH) insensitivity due to primary GH receptor deficiency. Endocr Rev 15: 369-390

Short RV (1994) Why sex? In: Short RV, Balaban E (eds) The differences between the sexes. Cambridge University Press, Cambridge, pp 3-22

Martin RD, Willner LA, Dettling A (1994) The evolution of sexual size dimorphism in primates. In: Short RV, Balaban E (eds) The d Tannenbaum GS, Martin JB (1976) Evidence for an endogenous ultradian rhythm governing growth hormone secretion in the rat. Endocrinology 98: 562-570

Tannenbaum GS, Choi HK, Gurd W, Waxman DJ (2001) Temporal relationship between the sexually dimorphic spontaneous GH secretory profiles and hepatic STAT5 activity. Endocrinology 142: 4599-4606

Teglund S, McKay C, Schuetz E, van Deursen JM, Stravopodis D, Wang D, Brown M, Bodner S, Grosveld G, Ihle JN (1998) Stat5a and Stat5b proteins have essential and nonessential, or redundant roles in cytokine responses. Cell 93: 841-850

Udy GB, Towers RP, Snell RG, Wilkins RJ, Park SH, Ram PA, Waxman DJ, Davey HW (1997) Requirement of STAT5b for sexual dimorphism of body growth rates and liver gene expression. Proc Natl Acad Sci USA 94: 7239-7244

Waxman DJ, Pampori NA, Ram PA, Agrawal AK, Shapiro BH (1991) Interpulse interval in circulating growth hormone patterns regulates sexually dimorphic expression of hepatic cytochrome P450. Proc Natl Acad Sci USA 88:6868-6872

Genetic Control of Size at Birth

D.B. Dunger[1], C.J. Petry[1] and K.K. Ong[1]

Summary

Size at birth is a heritable trait: estimates vary between 30% and 70%, but associations may be confounded by interactions with the maternal-uterine environment. Maternal smoking, length of gestation and parity could confound overall estimates of birth weight inheritance. However, the effects of maternal blood pressure, weight gain and glucose levels may be less easy to categorise, as they may reflect genetic factors acting in the mother.

First pregnancies are associated with apparent "restraint" of fetal growth. Offspring birth weight in these pregnancies is lower and correlates closely with the mother's own birth weight, possibly indicating a predominantly maternal inheritance. Candidate genes for maternal transmission of low birth weight include the mitochondrial DNA 16189 variant and exclusively maternally expressed genes such as *H19*. Offspring birth weight correlations with maternal blood pressure indicate that other maternal genes could influence size at birth in first pregnancies, where risk of pre-eclampsia is greatest. In subsequent pregnancies, gestational diabetes is linked to increased risk of macrosomia, and the impact of higher maternal glucose levels on larger offspring birth weight is demonstrated by a study of families with rare glucokinase gene mutations and in population studies of a common glucokinase gene promoter variant.

Larger birth weight shows a more autosomal mode of inheritance. Potential candidate genes reflect the importance of IGF-I, IGF-2, insulin and their respective receptors in regulating fetal growth, as shown by mouse knockout models and rare genetic variants in human subjects. Identification of common genetic variants associated with size at birth has been less successful. Birth weight association studies with polymorphisms in genes related to IGF-I, insulin and IGF-2 expression have yielded variable results. The common *INS* VNTR mini-satellite, which regulates *INS* and *IGF2* expression, has been associated with size at birth and is confirmed by parental allele transmission, but has not been replicated in all populations. Animal data indicate the important role of imprinted genes in fetal growth, possibly reflecting the conflict between maternal and paternal influences on size at birth and fetal survival. Although evidence in contemporary

---

[1] Department of Paediatrics, University of Cambridge, Addenbrooke's Hospital, Box 116 level 8, Hills Road, Cambridge CB2 2QQ, UK

Carel et al.
Deciphering Growth
© Springer-Verlag Berlin Heidelberg

Table 1. Significant maternal determinants of offspring birth weight. Results of a covariate model analysis in the ALSPAC children in focus cohort (n=1335).[a]

Factor	Standardised regression coefficient
Gestation	0.34
Sex of baby	-0.17
Parity	0.26
Mother's birth weight	0.23
Mother's weight	0.18
Mother's weight gain	0.10
Mother's height	0.12
Mother's smoking	-0.11

[a] Adjusted for offspring sex. Together, these factors explained 35% of the variance in offspring birth weight. Mother's age and education history (social status) did not contribute significantly to the model.

human populations remains elusive, preliminary data indicate that such models remain important for future study.

Introduction

Size at birth has been reported to be a highly heritable trait. Estimates of heritability from studies of monozygous and dizygous twins range from 30 to 70% (Magnus 1984; Little and Sing 1987). However, these estimates of heritability could be confounded by the profound effects of maternal-uterine environment on size at birth. Furthermore, from family studies, Ounsted et al. (1988) reported an apparently stronger relationship between maternal and offspring birth weights among smaller-born infants than in larger birth weight offspring.

In a contemporary birth cohort, the Avon Longitudinal Study of Parents and Children (ALSPAC; Golding et al. 2001), size at birth was significantly related to a number of maternal factors, including parity, gestation, mother's adult size, and mother's own birth weight (Table 1). Maternal under-nutrition and disease are less common determinants of birth weight in contemporary populations (Mathews et al. 1999). Maternal hypertension and pre-eclampsia are associated with impaired placental function and reduced birth weight, whereas maternal diabetes often results in fetal macrosomia. Thus, size at birth is a complex trait that reflects both the maternal-uterine environment and fetal genes.

Maternal uterine environment

Size at birth is the most important determinant of perinatal survival (Karn and Penrose 1951), yet, in most populations, mean birth weight is slightly lower than that which is optimal for offspring survival (Alberman 1991). Studies have led to the concept that fetal growth is usually subject to some degree of maternal-uterine restraint (Gluckman et al. 1990). The genetic growth potential of the fetus will put nutritional demands on the mother, which could threaten her survival at times of poor nutrition. Haig (1996) pointed out the inherent conflict between maternal and fetal survival, leading to the hypothesis that imprinted genes evolved to reflect the competing interests of mother and father. Imprinting results in silencing of either the maternal or the paternal copy of a gene, and thus exclusive expression of the allele inherited from one of the parents (Reik and Walter 2001). Animal data support this hypothesis, in that the protein product of the paternally expressed *IGF2* gene promotes fetal growth, whereas the gene for the IGF-2 receptor (*IGF2R*), which clears IGF-2 from the circulation and reduces fetal growth, is exclusively maternally expressed (Ludwig et al. 1996). These hypotheses have been developed in animal models and may reflect the situation in species where there is large litter size and where even the postnatal growth potential of the offspring may affect maternal survival through demands on lactation (Haig and Wilkins 2000). The degree to which similar selection pressure has affected human genetic determinants of fetal growth is uncertain. Although some parallels can be seen with the animal models, there are differences that may be critical to successful human pregnancy. Singleton births are usual in humans, and in most populations multiple pregnancies are relatively rare. The factor that may be unique to human pregnancies is the ability of the mother to deliver an offspring with a large head size.

In human offspring, restraint of fetal growth is most evident in first pregnancies. The evidence for this comes from the observation that offspring of first pregnancies have a lower birth weight, thinner birth size, and relatively preserved head circumference and length, suggesting reduced adiposity at birth; they also demonstrate rapid postnatal catch-up growth (Ong et al. 2002a). Postnatal catch-up growth may be an important marker of prenatal growth restraint, as it is predicted by maternal factors such as first pregnancies, maternal smoking in pregnancy and mother's own low birth weight (Ong et al. 2000a). It is also evident in infants whose intra-uterine environment has been affected by poor placental function, for example, secondary to maternal hypertension and pre-eclampsia (PET).

It is in such smaller, or growth-constrained, infants that Ounsted et al. (1988) reported stronger association between offspring birth weight and maternal birth weight. These associations are also reflected in an increased risk for PET in first pregnancies. In a recent study of over 4,000 pregnancies in Cambridge (UK), we noted that mothers' first pregnancies were associated with a lower offspring birth weight (mean difference = 130 g, P<0.0001) and increased risk of pregnancy-induced hypertension (OR = 4.3, 95% CI: 2.5 to 7.3; P<0.0001). Thus the mechanism of fetal growth restraint in first pregnancies may also be linked to risk of pregnancy-induced hypertension. The restraint of fetal growth may occur through inhibition of spiral artery invasion of the uterine wall during early conception, and this site has also been implicated as the origin of circulating factors implicated in the pathogenesis of pre-eclampsia. Lower maternal birth weight has been related to both increased risk of pre-eclampsia (Innes et al. 2003) and greater restraint of fetal growth resulting in small babies (Ounsted et al. 1986). In our own studies of the ALSPAC cohort, postnatal catch-up growth was more common in first pregnancies and it was in those pregnancies that the strongest relationships with maternal birth weight were observed.

Maternal genes and size at birth

Inheritance through the maternal line of restraint of fetal growth and reduced birth size could involve mitochondrial genes, which are exclusively transmitted from the mother. The mitochondrial DNA 16189 variant has been reported by our group to be associated with thinner offspring size at birth (Casteels et al. 1999). Infants with the 16189 variant showed increased postnatal weight gain, suggesting that the effects might be mediated by the maternal uterine environment, although the actual mechanisms have not been clarified. Another potential mechanism whereby restraint of fetal growth could be transmitted through the maternal line is inheritance of exclusively maternally expressed genes, where paternal alleles are silenced by imprinting. We have recently identified association with size at birth and a common variation in a maternally expressed gene *H19*, which regulates imprinting of the exclusively paternally expressed fetal growth promoter *IGF2*. The associations were seen in two independent birth cohorts. Due to the relatively lower number of complete, informative parent-offspring trios in our study, it was not possible to distinguish whether the birth size association was directly due to inheritance of the maternal allele or indirectly from effects of mother's genotype on the uterine environment. The *H19* variant was further associated with cord blood IGF-2 protein levels and also with maternal glucose levels during pregnancy, suggesting a potential complex interaction between maternal and fetal genotypes. Further complexity of such interactions has recently been demonstrated by the identification of placental specific promoter *IGF2* expression (Constancia et al. 2002). Knock-out of that regulator leads to initial compensatory up-regulation of placental nutrient transfer, which could have effects on maternal metabolic status, but subsequently there is failure of compensatory nutrient transfer, fetal growth slows and birth weights are low (Sibley et al. 2004).

The associations between *H19*, size at birth and maternal glucose levels were particularly seen in first pregnancies. Interestingly, the relationship between size at birth and maternal glucose levels at 28 weeks gestation is altered in first pregnancies (Fig. 1). Although the overall relationship between maternal glucose levels and offspring birth weight is evident, glucose levels are higher in first pregnancies than in subsequent pregnancies. After the first pregnancy, the risk of gestational diabetes increases, and we observed that maternal glucose levels steadily increased again from second to third and subsequent pregnancies. The inferred mechanism is that increased glucose levels in the mother leads to in-

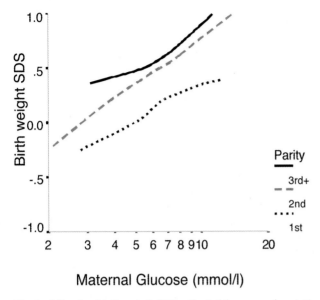

Fig. 1. Offspring birth weigth SDS, adjusted for sex and gestational age, related to mother's stimulated plasma glucose levels at 28 weeks gestation, and divided by parity (birth order). Data from 3,000 Rosie Maternity Hospital, Cambridge

creased glucose transfer to the offspring and fetal β-cell hyperplasia, increased insulin secretion, and greater fetal adiposity. However, effects on offspring birth weight may also be seen with more subtle increases in maternal glucose. Study of mothers with rare genetic defects of the glucokinase gene, which result in increased maternal glucose levels, confirm that increases in glucose transport across the placenta can result in larger size at birth (Hattersley et al. 1998). A recent study reports that common variation in the glucokinase gene promoter also relates to size at birth (Weedon et al. 2005). In our own studies in normal pregnancies, a continuous relationship is observed between maternal glucose levels and offspring birth weight (Fig. 1). Lower offspring birth weight is also associated with lower fasting maternal insulin levels, indicating that increased insulin sensitivity in the mother may result in reduced transfer of nutrients across the placenta. Higher maternal insulin secretion post-oral glucose load was also associated with lower birth weight, presumably through reducing maternal post-prandial glucose levels. These data indicate that within the normal range of birth weight, variation in mothers' glucose levels, insulin sensitivity and insulin secretion, due to maternal weight gain or maternal genes, may have an important influence on size at birth. Such maternal genetic influences may differ between first pregnancies, where birth weight inheritance may be more maternally transmitted, compared to subsequent pregnancies. In second and subsequent pregnancies, the effects of genes expressed from both parents, or exclusively from paternal genes, may be more evident.

Fetal genes and size at birth

Over the last 10 years, a series of elegant animal knock-out experiments have identified the importance of IGF-1, IGF-2, insulin and their respective receptors in regulating fetal growth and size at birth (Baker et al. 1993; Fig. 2). IGF-2 may be important for early fetal growth, acting through the type 1 IGF receptor, whereas IGF-1 may be a more important determinant of later fetal growth. Insulin, by acting through the insulin receptor, may be an important regulator of fetal adiposity and may also promote fetal growth through its permissive effect on IGF-1 generation at the liver.

Studies of rare mutations in human subjects support the role of these peptides and receptors in the regulation of human fetal growth. Newborns with defects in pancreatic development or insulin receptor activation show reduced fetal growth and adiposity (Ogilvy-Stuart et al. 2001). Although only two subjects have been reported with defects of the *IGF1* gene, both were very small at birth, with a particular reduction in head size (Woods et al. 1996). Recently, infants born small for gestational age (SGA), reflecting severe intrauterine growth retardation, had been reported with genetic defects in the type 1 IGF receptor (Abuzzahab et al. 2003). Infants with a reduced or increased copy number of the type 1 IGF receptor were also reported to show reduced and increased fetal and postnatal growth, respectively (Okubo et al. 2003), indicating that the copy number of IGF1R may influence growth in humans. Subjects with the fetal overgrowth Beckwith Wiedemann syndrome had over-expression of IGF-2 in association with genetic defects in the *IGF2* gene (Morison et al. 1996). As yet, there have been no reported human cases with severe mutations of the *IGF2R* gene, but variable expression of this gene has been reported in relation to size at birth (Wutz et al. 1998).

In population studies, cord blood measurements of IGF-1, IGF-2 and insulin have been used as surrogate measures of the activity of these peptides in regulating fetal growth. All three peptides have been shown to be positively related to size at birth (Ong et al. 2000b). In contrast, growth hormone levels tend to be high in babies born SGA, perhaps reflecting the metabolic rather than the anabolic role of growth hormone in the perinatal period (Ogilvy-Stuart et al. 1998). Levels of the soluble form of the IGF-2 receptor, and higher levels of IGF2R relative to IGF-2, are negatively associated with size at birth and placental weight (Ong et al. 2002b), suggesting that the IGF-2 regulatory and growth inhibitory functions of this receptor observed in mouse and in vitro models (Ludwig et al. 1996) also apply to humans.

The extent to which common genetic polymorphisms are related to human size at birth is uncertain, and the current data are limited. Polymorphism of the *IGF1* gene has been explored in a number of studies, and the same variants have been shown to be variably associated with size at birth. The common *IGF1* promoter CA repeat has been reported to be associated with size at birth (Vaessen et al. 2002; Johnston et al. 2003), however, this association was not confirmed in a large population study (Frayling et al. 2002). The discrepancy between these studies may relate to the selection of subjects. Studies of populations of children born SGA and with postnatal short-stature could over-select for certain genetic variants associated with size at birth.

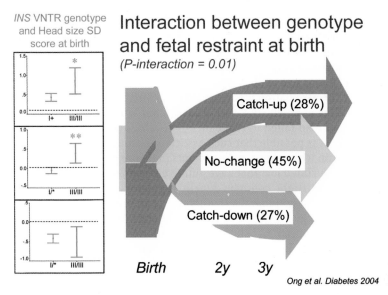

Fig. 2. Association between size at birth and *INS* VNTR genotype in the ALSPAC children in focus cohort: significant interaction was seen with rapid "catch-up" postnatal weight gain, a marker of in utero growth restraint. (From Ong et al. 2004)

Recently, we examined the relationship between common variation in *IGF2* and *IGF2R* and size at birth in a large representative birth cohort. Two *IGF2* SNPs were selected on the basis of their reported association with body mass index (BMI) in middle-aged men (ApaI and +6815; Gaunt 2001). The ApaI SNP was significantly associated with BMI in children at age seven but did not show any consistent association with size at birth. The +6815 SNP was weakly associated with birth length but not with birth weight or head circumference. Neither was there evidence of any parent-of-origin effects for either of these *IGF2* SNPs. For *IGF2R*, we chose to study a polymorphism that causes a non-conservative glycine to arginine amino acid change in the receptor located in its IGF-2 binding region (Killian et al. 2001). Again, while this variant was related to height gains over the first two years of life, we did not observe association with size at birth and there was no evidence of parent-of-origin effects.

In contrast to these largely negative findings, the common *INS* VNTR polymorphism, which regulates both *INS* and *IGF2* transcription (Bennett et al. 1996; Paquette et al. 1998), has been associated with size at birth. We originally reported association between the *INS* VNTR class III/III genotype and larger size at birth, particularly in relation to head circumference and, to a lesser extent, with length and birth weight (Dunger et al. 1998). We have subsequently confirmed these associations with head circumference at birth by association in a second cohort, and also by parental class III allele transmission, excluding potential confounding of the association by population stratification (Ong et al. 2004). Interestingly, these observations were strongest in second and subsequent pregnancies, where potential confounding of fetal genetic effects by maternal restraint of fetal growth is less common. Consistent with the birth size association and the biological basis, the *INS* VNTR class III/III genotype was also associated with higher cord blood IGF-2 levels (Ong et al. 2004). Curiously, association between the *INS* VNTR and size at birth was also observed in the Pima Indians of Arizona, USA, but the association was reversed and class III subjects had lower birth weights (Lindsay et al. 2003). Other studies from Finland and the southwest of England have also failed to replicate our findings (Bennett et al. 2004; Mitchell et al. 2004). It is possible that such associations may be confounded by linkage disequilibrium with other genetic variants in this region of chromosome 11, which is rich in imprinted loci that have putative effects on fetal growth.

The impetus to identify the common genetic regulators of size at birth has increased with the observation that the size at birth is an important predictor of adult disease risk (Hales and Barker 2001). Genetic determinants of obesity and adult metabolic syndrome risk have been examined for association with size at birth. Our own group have looked at common variations in *IRS1* and *PPARγ* as both have been associated with risk of insulin resistance in adult populations. However, we were unable to confirm any association with size at birth (Mason et al. 2000). Common variation of the *ACE* gene is related to hypertension and cardiovascular disease risk in adults, and association was observed with size at birth in the ALSPAC cohort (unpublished observations). The mechanism underlying such association is unclear, but *ACE* is known to regulate metabolism and could influence placental function. Common polymorphism in the G protein beta3 subunit gene has been associated with low birth weight in pregnancies without other risks for reduced fetal growth (Hocher et al. 2000). Other common genetic variants reported to be associated with size at birth include angiotensinogen (Zhang et al. 2003), the small heterodimer partner in a cohort of obese children (Nishigori et al. 2001), phosphoglucomutase locus 1 in girls (Gloria-Bottini et al. 2001), and the vitamin D receptor (Lorentzon et al. 2000), preproneuropeptide Y (Karvonen et al. 2000), and acid phosphatase in boys (Amante et al. 1990).

Further genetic polymorphisms in placental alkaline phosphatase in the fetus (Magrini et al. 2003), and maternal aromatic compound-inducible cytochrome P450 and glutathione-S-transferase genes (Wang et al. 2002) were associated with modifications of the effects of maternal smoking during pregnancy on offspring birth weight. Other maternal genetic polymorphisms that may influence maternal metabolism are reported to be associated with size at birth include methlenetetrahydrofolate reductase (Nurk et al. 2004) and G protein beta3 subunit (Masuda et al. 2002).

Conclusions

Size at birth is a critical determinant of perinatal survival and must have been subject to intense selection pressure during human history. The Haig hypothesis (1996) concerning the conflicting interests between the mother and father on offspring growth may be less important with respect to human pregnancies, where multiple pregnancy is the exception and the successful delivery of a large fetal head size is critical to maternal survival. Nevertheless, imprinted genes may still have an important role in regulating size at birth. It is estimated that 70% of all known imprinted genes may regulate fetal growth and brain development (Reik and Walter 2001).

Identification of common genetic variants that influence size at birth will require consideration of complex paradigms that include effects of the maternal uterine environment and the potential influence of the maternal genotype on that environment. Different rates of transition from conditions of poor nutrition and low perinatal survival to abundant nutrition and vastly improved perinatal mortality could have variable effects on gene frequencies in different populations and could alter the interaction between maternal genotype and the maternal-uterine environment. For example, in populations that have adapted to many generations of nutritional deprivation, very rapid increases in nutrition and risk of obesity could switch a maternal-uterine environment that favours restraint of fetal growth to one enhancing the risk of gestational diabetes and fetal macrosomia.

Studies to identify common genetic variations associated with size at birth have adopted several strategies, which have their advantages and disadvantages. One common approach has been to utilise growth clinic collections of children born SGA who remain short during childhood. However, as over 90% of SGA children show postnatal catch-up growth, this approach selects children who are largely characterised by failure of postnatal catch-up gains in weight and height. Such infants are likely to be a highly heterogeneous group, with potential novel defects of skeletal development or as yet unrecognised defects of the growth hormone-IGF-1 axis that restrict postnatal catch-up. Identification of such variants in short SGA children may reveal important new therapeutic targets and modalities; however, they are unlikely to increase our understanding of the regulation of normal fetal growth.

In contrast, population studies with good ascertainment are more likely to inform identification of common genetic determinants of size at birth. Such studies require more resources and are more expensive and, from our arguments detailed above, these cohorts should include a complex phenotype. As well as accurate measurement of size at birth, including weight, length, and head circumference, these studies would be helped by inclusion of some measure of postnatal weight gain over the first year of life. These studies would be further enriched by including maternal pregnancy data, such as maternal blood pressure and glucose levels, which could affect size at birth. Genetic assessments should not be confined to offspring genotypes but should also include maternal genotypes related to potential effects on maternal metabolism and the uterine environment. Assessment of both parental genotypes may allow analysis of allele transmission and parent-of-origin effects. These analyses are often subject to problems of poor statistical power, and very large sample sizes may be required to differentiate effects of maternal genotype from those of maternal allele transmission on size at birth. Such large comprehensive genetic studies are rare, but they may be necessary to unravel the reported links between size at birth and adult disease.

References

Abuzzahab MJ, Schneider A, Goddard A, Grigorescu F, Lautier C, Keller E, Kiess W, Klammt J, Kratzsch J, Osgood D, Pfaffle R, Raile K, Seidel B, Smith RJ, Chernausek SD (2003) IGF-I receptor mutations resulting in intrauterine and postnatal growth retardation. N Engl J Med 349: 2211-2222.

Alberman E (1991) Are our babies becoming bigger? J R Soc Med 84: 257-260.

Amante A, Gloria-Bottini F, Bottini E (1990) Intrauterine growth: association with acid phosphatase genetic polymorphism. J Perinat Med 18: 275-282.

Baker J, Liu JP, Robertson EJ, Efstratiadis A (1993) Role of insulin-like growth factors in embryonic and postnatal growth. Cell 75: 73-82.

Bennett AJ, Sovio U, Ruokonen A, Martikainen H, Pouta A, Taponen S, Hartikainen AL, King VJ, Elliott P, Jarvelin MR, McCarthy MI (2004) Variation at the insulin gene VNTR (variable number tandem repeat) polymorphism and early growth: studies in a large Finnish birth cohort. Diabetes 53: 2126-2131.

Bennett ST, Wilson AJ, Cucca F, Nerup J, Pociot F, McKinney PA, Barnett AH, Bain SC, Todd JA (1996) IDDM2-VNTR-encoded susceptibility to type 1 diabetes: dominant protection and parental transmission of alleles of the insulin gene-linked minisatellite locus. J Autoimmun 9: 415-421.

Casteels K, Ong KK, Phillips DI, Bednarz A, Bendall H, Woods KA, Sherriff A, the ALSPAC Study Team, Golding J, Pembrey ME, Poulton J, Dunger DB (1999) Mitochondrial 16189 variant, thinness at birth and type 2 diabetes. Lancet 353: 1499-1500.

Constancia M, Hemberger M, Hughes J, Dean W, Ferguson-Smith A, Fundele R, Stewart F, Kelsey G, Fowden A, Sibley C, Reik W (2002) Placental-specific IGF-II is a major modulator of placental and fetal growth. Nature 417: 945-948.

Dunger DB, Ong, KK Huxtable SJ, Sherriff A, Woods KA, Ahmed ML, Golding J, Pembrey ME, Ring S, Bennett ST, Todd JA (1998) Association of the INS VNTR with size at birth. Nature Genet 19: 98-100.

Frayling TM, Hattersley AT, McCarthy A, Holly J, Mitchell SM, Gloyn AL, Owen K, Davies D, Smith GD, Ben-Shlomo Y (2002). A putative functional polymorphism in the IGF-I gene: association studies with type 2 diabetes, adult height, glucose tolerance, and fetal growth in UK populations. Diabetes 51: 2313-2316.

Gaunt TR, Cooper JA, Miller GJ, Day IN, O'Dell SD (2001) Positive associations between single nucleotide polymorphisms in the IGF2 gene region and body mass index in adult males. Human Mol Genet10: 1491-1501.

Gloria-Bottini F, Lucarini N, Palmarino R, La Torre M, Nicotra M, Borgiani P, Cosmi E, Bottini E (2001) Phosphoglucomutase genetic polymorphism of newborns. Am J Hum Biol 13: 9-14.

Gluckman PD, Breier BH, Oliver M, Harding J, Bassett N (1990) Fetal growth in late gestation – a constrained pattern of growth. Acta Paediatr Scand Suppl 367: 105-110.

Golding J, Pembrey ME, Jones R (2001) ALSPAC – the Avon Longitudinal Study of Parents and Children. I. Study methodology. Paed Perinat Epidemiol 15: 74-87.

Haig D (1996) Altercation of generations: genetic conflicts of pregnancy. Am J Reprod Immunol. 35: 226-232.

Haig D, Wilkins JF (2000) Genomic imprinting, sibling solidarity and the logic of collective action. Philos Trans R Soc Lond B Biol Sci 355: 1593-1597.

Hales CN, DJ Barker (2001) The thrifty phenotype hypothesis. Brit Med Bull 60: 5-20.

Hattersley AT, Beards F, Ballantyne E, Appleton M, Harve R Ellard S (1998) Mutations in the glucokinase gene of the fetus result in reduced birth weight. Nature Genet 19: 268-270.

Hocher B, Slowinski T, Stolze T, Pleschka A, Neumayer HH, Halle H (2000) Association of maternal G protein beta3 subunit 825T allele with low birthweight. Lancet 355: 1241-1242.

Innes KE, Byers TE, Marshall JA, Baron A, Orleans M, Hamman RF (2003) Association of a woman's own birth weight with her subsequent risk for pregnancy-induced hypertension. Am J Epidemiol 158: 861-870.

Johnston LB, Dahlgren J, Leger J, Gelander L, Savage MO, Czernichow P, Wikland KA, Clark AJ (2003) Association between insulin-like growth factor I (IGF-I) polymorphisms, circulating IGF-I, and pre- and postnatal growth in two European small for gestational age populations. J Clin Endocrinol Metab 88: 4805-4810.

Karn MN, LS Penrose (1951) Birth weight and gestation time in relation to maternal age, parity and infant survival. Ann Eugen 16: 147-158.

Karvonen MK, Koulu M, Pesonen U, Uusitupa MI, Tammi A, Viikari J, O Simell, Ronnemaa T (2000) Leucine 7 to proline 7 polymorphism in the preproneuropeptide Y is associated with birth weight and serum triglyceride concentration in preschool aged children. J Clin Endocrinol Metab 85: 1455-1460.

Killian JK, Oka Y, Jang HS, Fu X, Waterland RA, Sohda T, Sakaguchi S, Jirtle RL (2001) Mannose 6-phosphate/insulin-like growth factor 2 receptor (M6P/IGF2R) variants in American and Japanese populations. Human Mutat 18: 25-31.

Lindsay RS, Hanson RL, Wiedrich C, Knowler WC, Bennett PH, Baier LJ (2003) The insulin gene variable number tandem repeat class I/III polymorphism is in linkage disequilibrium with birth weight but not Type 2 diabetes in the Pima population. Diabetes 52: 187-193.

Little RE, Sing CF (1987) Genetic and environmental influences on human birth weight. Am J Human Genet 40: 512-526.

Lorentzon M, Lorentzon R, Nordstrom P (2000) Vitamin D receptor gene polymorphism is associated with birth height, growth to adolescence, and adult stature in healthy caucasian men: a cross-sectional and longitudinal study. J Clin Endocrinol Metab 85: 1666-1670.

Ludwig T, Eggenschwiler J, Fisher P, D'Ercole ML, Efstratiadis A (1996) Mouse mutants lacking the type 2 IGF receptor (IGF2R) are rescued from perinatal lethality in Igf2 and Igf1r null backgrounds. Dev. Biol. 177: 517-535.

Magnus P (1984) Causes of variation in birth weight: a study of offspring of twins. Clin Genet 25: 15-24.

Magrini A, Bottini N, Gloria-Bottini F, Stefanini L, Bergamaschi A, Cosmi E, Bottini E (2003) Enzyme polymorphisms, smoking, and human reproduction. A study of human placental alkaline phosphatase. Am J Hum Biol 15: 781-5.

Mathews F, Yudkin P, Neil A (1999) Influence of maternal nutrition on outcome of pregnancy: prospective cohort study. Brit Med J 319: 339-343.

Mason S, Ong KK, Pembrey ME, Woods KA, Dunger DB (2000) The Gly972Arg variant in insulin receptor substrate-1 is not associated with birth weight in contemporary English children. The ALSPAC Study Team. Avon Longitudinal Study of Pregnancy and Childhood. Diabetologia 43: 1201-1202.

Masuda K, Osada H, Iitsuka Y, Seki K, Sekiya S (2002) Positive association of maternal G protein beta3 subunit 825T allele with reduced head circumference at birth. Pediatr Res 52: 687-691.

Mitchell SM, Hattersley AT, Knight B, Turner T, Metcalf BS, Voss LD, Davies D, McCarthy A, Wilkin TJ, Smith GD, Ben-Shlomo Y, Frayling TM (2004) Lack of support for a role of the insulin gene variable number of tandem repeats minisatellite (INS-VNTR) locus in fetal growth or type 2 diabetes-related intermediate traits in United Kingdom populations. J Clin Endocrinol Metab 89: 310-317.

Morison IM, Becroft DM, Taniguchi T, Woods CG, Reeve AE (1996) Somatic overgrowth associated with overexpression of insulin- like growth factor II. Nature Med 2: 311-316.

Nishigori H, Tomura H, Tonooka N, Kanamori M, Yamada S, Sho K, Inoue I, Kikuchi N, Onigata K, Kojima I, Kohama T, Yamagata K, Yang Q, Matsuzawa Y, Miki T, Seino S, Kim MY, Choi HS, Lee YK, Moore DD, Takeda J (2001) Mutations in the small heterodimer partner gene are associated with mild obesity in Japanese subjects. Proc Natl Acad Sci USA 98: 575-580.

Nurk E, Tell GS, Refsum H, Ueland PM, Vollset SE (2004) Associations between maternal methylenetetrahydrofolate reductase polymorphisms and adverse outcomes of pregnancy: the Hordaland Homocysteine Study. Am J Med 117: 26-31.

Ogilvy Stuart AL, Hands SJ, Adcock CJ, Holly JM, Matthews DR, Mohamed Ali V, Yudkin JS, Wilkinson AR, Dunger DB (1998) Insulin, insulin-like growth factor I (IGF-I), IGF-binding protein-1, growth hormone, and feeding in the newborn. J Clin Endocrinology Metab 83: 3550-3557.

Ogilvy-Stuart AL, Soos MA, Hands SJ, Anthony MY, Dunger DB, O'Rahilly S (2001) Hypoglycemia and resistance to ketoacidosis in a subject without functional insulin receptors. J Clin Endocrinol Metab 86: 3319-3326.

Okubo Y, Siddle K, Firth H, O'Rahilly S, Wilson LC, Willatt L, Fukushima T, Takahashi S, Petry CJ, Saukkonen T, Stanhope R, Dunger DB (2003) Cell proliferation activities on skin fibroblasts from a short child with absence of one copy of the type 1 insulin-like growth factor receptor (IGF1R) gene and a tall child with three copies of the IGF1R gene. J Clin Endocrinol Metab 88: 5981-5988.

Ong KK, Ahmed ML, Emmett PM, Preece MA, Dunger DB, the ALSPAC Study Team (2000a) Association between postnatal catch-up growth and obesity in childhood: prospective cohort study. Brit Med J 320: 967-971.

Ong KK, Kratzsch J, Kiess W, the ALSPAC Study Team, M Costello, CD Scott, DB Dunger (2000b) Size at birth and cord blood levels of insulin, insulin-like growth factor I (IGF-I), IGF-II, IGF-binding protein-1 (IGFBP-1), IGFBP-3, and the soluble IGF-II/mannose-6-phosphate receptor in term human infants. J Clin Endocrinol Metab 85: 4266-4269.

Ong KK, Preece MA, Emmett PM, Ahmed ML, Dunger DB (2002a) Size at birth and early childhood growth in relation to maternal smoking, parity and infant breast-feeding: longitudinal birth cohort study and analysis. Pediat Res 52: 863-867.

Ong KK, Kratzsch J, Kiess W, the ALSPAC Study Team, Dunger DB (2002b) Circulating IGF-I levels in childhood are related to both current body composition and early postnatal growth rate. J Clin Endocrinol Metab 87: 1041-1044.

Ong KK, Petry CJ, Barratt BJ, Ring S, Cordell HJ, Wingate DL, Pembrey ME, Todd JA, Dunger DB (2004) Maternal-fetal interactions and birth order influence insulin variable number of tandem repeats allele class associations with head size at birth and childhood weight gain. Diabetes 53: 1128-1133.

Ounsted M, Scott A, Moar VA (1988) Constrained and unconstrained fetal growth: associations with some biological and pathological factors. Ann Human Biol 15: 119-129.

Ounsted M, Scott A, Ounsted C (1986) Transmission through the female line of a mechanism constraining human fetal growth. Ann Human Biol 13: 143-151.

Paquette J, Giannoukakis N, Polychronakos C, Vafiadis P, Deal C (1998) The INS 5' variable number of tandem repeats is associated with IGF2 expression in humans. J Biol Chem 273: 14158-14164.

Reik W, Walter J (2001) Genomic imprinting: parental influence on the genome. Nature Rev Genet 2: 21-32.

Sibley CP, Coan PM, Ferguson-Smith AC, Dean W, Hughes J, Smith P, Reik W, Burton GJ, Fowden AL, Constancia M (2004) Placental-specific insulin-like growth factor 2 (Igf2) regulates the diffusional exchange characteristics of the mouse placenta. Proc Natl Acad Sci USA 101: 8204-8208.

Vaessen N, Janssen JA, Heutink P, Hofman A, Lamberts SW, Oostra BA, Pols HA. van Duijn CM (2002) Association between genetic variation in the gene for insulin-like growth factor-I and low birthweight. Lancet 359: 1036-1037.

Wang X, Zuckerman B, Pearson C, Kaufman G, Chen C, Wang G, Niu T, Wise PH, Bauchner H, Xu X (2002) Maternal cigarette smoking, metabolic gene polymorphism, and infant birth weight. JAMA 287: 195-202.

Weedon MN, Frayling TM, Shields B, Knight B, Turner T, Metcalf BS, Voss L, Wilkin TJ, McCarthy A, Ben-Schlomo Y, Davey Smith G, Ring S, Jones R, Golding J, Byberg L, Mann V, Axelsson T, Syvanen AC, Leon D, Hattersley AT (2005) Genetic regulations of birth weight ans fasting glucose by a common polymorphism in the islet cell promoter of the glucoki-nase gene. Diabetes. 54:576-581

Woods KA, Camacho Hubner C, Savage MO, Clark AJ (1996) Intrauterine growth retardation and postnatal growth failure associated with deletion of the insulin-like growth factor I gene. New Engl J Med 335: 1363-1367.

Wutz A, Smrzka OW, Barlow DP (1998)Making sense of imprinting the mouse and human IGF2R loci. Novartis Foundation Symposia 214: 251-259.

Zhang XQ, Varner M, Dizon-Townson D, Song F, Ward K (2003) A molecular variant of angio-tensinogen is associated with idiopathic intrauterine growth restriction. Obstet Gynecol 101: 237-242.

The GH/IGF-1 Axis: Insights from Animal Models

Martin Holzenberger[1], Laurent Kappeler[1],
Carlos De Magalhaes Filho[1] and Yves Le Bouc[1]

Summary

Individuals develop from single cells through a genetically controlled program that regulates cell growth, cell proliferation and differentiation. The quantitative equilibrium between cell differentiation and proliferation is particularly important for tissue-specific growth and the shaping of higher organisms. Insulin-like growth factors (IGF) are key regulators of somatic growth, and growth hormone (GH), by controlling important aspects of IGF activity in many tissues in mammals, is able to coordinate this growth in a defined, spatio-temporal manner at the whole body level. Using homologous recombination, we generated mouse models with genetically determined IGF-1R insufficiency. We showed that partial inactivation of IGF-1R causes postnatal growth deficits that appear during the postnatal growth spurt and persist in the adult. We found that these growth deficits depend on the dosage of the IGF-1R gene. In our mutant mice, the postnatal growth of males relied more strongly on IGF-1R levels than the growth of females. Experiments using tissue-specific IGF-1R inactivation in the central nervous system provided evidence that IGF signaling in the brain may play a key role during the development of the somatotrope function in mammals.

Introduction

Growth is among the most fascinating aspects of biology. In this context and in very simple words, growth is the process by which individual organisms increase their body mass until they reach their adult size. The study of growth tries to elucidate the biological mechanisms through which individuals are able to increase body size. Historically, at an early time point of understanding of (human) reproduction, it was believed that complete body plans were already present in sperm and that growth consisted of providing resources to the preexisting homunculus, the task being to simply enlarge its frame. Retrospectively, such a hypothesis is, of course, incompatible with almost any finding of modern biology (e.g., there was no explanation of how the following generations were

[1] Inserm U515, Hôpital Saint-Antoine, 75571 Paris 12, France

Carel et al.
Deciphering Growth
© Springer-Verlag Berlin Heidelberg

preformed within their ancestors). Mentioning this hypothesis should simply underscore that growth cannot just be a linear process; it must comprise many elaborate mechanisms that control the different stages of development to provide, at any moment of embryogenesis and during all later developmental periods, the appropriate growth-promoting signals. Together with the discovery of the cellular basis of life, the early phases of individual development and growth were recognized to consist primarily of developmental processes during which the principle task lay in regulating the generation of transitory cell lineages from precursors and in regulating the fate of these cells, particularly to make the continuous decision of whether to maintain and further differentiate cell lineages or to abandon them to apoptosis. Instead of a homunculus, the single cell was the beginning of all individual life, and modern biology revealed irrefutably that body plans are the result of complex biochemical reactions, governed by large sets of genes that enable cells in different early embryonic structures to develop into specialized, differentiated cells.

Thus, development has become an issue of cell growth, cell proliferation and differentiation. In lower organisms, like C. elegans for example, the body plan is relatively simple and the number of cells in different body compartments is invariable. However, during evolution, many organisms quickly learned to shape their tissues by gaining control over cell proliferation during the different segments of development and by quantitatively regulating the equilibrium between cell differentiation and proliferation. In vertebrate ontogenesis, these mechanisms governing tissue growth apparently became of prime importance once individual organogenesis started. A large number of growth factors then play key roles in controlling the balance between proliferation and differentiation in the literally thousands of cell lineages composing the different organs and tissues of mammalian bodies. Through this control, bodies are shaped in a species-specific manner and overall body size is finally determined. IGF signaling has been recognized as one of the major molecular regulators of cell growth and proliferation (Nakae et al. 2001). Moreover, it is generally accepted that GH, by controlling important aspects of IGF activity in many tissues and cell types of mammals, is able to coordinate somatic growth in a defined spatio-temporal manner at the whole body level (Lupu et al. 2001). IGF signaling, however, not only regulates growth but also affects differentiation and may, through epigenetic processes, steer adult cell function as a result of particular conditions during postnatal development (Murakami et al. 2003). Much of our present understanding of growth regulation has been deduced from human growth and growth-related pathologies described in man (Denley et al. 2004; Abbuzahab et al. 2003; Woods et al. 1996). Moreover, over the last two decades, numerous mouse models have been developed and studied in detail, and they have provided us with valuable information about the genes that control mammalian growth (Dupont and Holzenberger 2003).

It appears through many of these studies that mammalian growth is a highly plastic process. Not only scientific results but also everyday life tells us this. Probably the most eloquent example for human growth plasticity is the so-called secular growth trend that started in the developed countries over a century ago (Fig. 1; Holzenberger et al. 1991). For many decades now, since the 19th century,

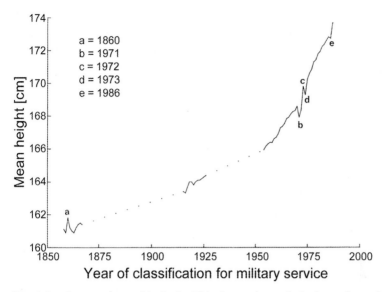

Fig. 1. Secular growth trend in Spain. This time series analysis shows the evolution of the mean body height of Spanish men between 1858 and 1987. In the 19th century and from 1971 to 1987, height was measured at age 18. From the beginning of the 20th century until 1970, however, height was measured at age 20. The letters **a** to **e** indicate the years when the procedures of recruitment were slightly modified. Interrupted lines indicate time periods for which no data were available (Modified from Holzenberger et al. 1991).

the average body size of humans has increased considerably. It is well accepted that better environment, nutrition, sanitary and health conditions are in large part responsible for this phenomenon; meanwhile genetic drift and the possibility that tall people generate more offspring than small people could be ruled out as explanations. Of course, the plasticity of mammalian growth needs a physiological, cellular and molecular basis, and I will try to condense some evidence pointing to a potential mechanism that could implement this plasticity in mammals.

Insulin-like pathways control growth

Homologs of insulin-like signaling pathways can be found in very simple organisms, and research from the last few decades has shown that insulin/IGF signaling cascades and their functions are highly conserved throughout eukaryote phylogenesis (Guarente and Kenyon 2000; Kenyon 2001; Gems and Partridge 2001). However, the number of different proteins and their genes that constitute these cascades is less conserved. The number of potential ligands, for example, varies from about 40 in C. elegans to only seven in Drosophila. In mammals, three essential ligands – IGF-I, IGF-II and insulin – have been identified and extensively characterized, but genes coding for several other structurally related peptides have been found, too. Typical for the evolution of the insulin/IGF system is the increased complexity of the extracellular part of the signaling network, probably through gene duplications of the main tyrosine kinase transmembrane receptor. This receptor is unique in worms and flies, with the respective proteins being DAF-2 isolated from C. elegans and INR, its homolog from Drosophila. During vertebrate evolution, however, this pathway seems to have divided into an IGF signaling pathway and an insulin pathway, with two separate transmembrane receptors, each of which is activated by its specific ligands (Ullrich et al. 1985; 1986). However, similarities in overall receptor structure and in ligand-binding characteristics between the insulin and the IGF system ensure molecular crosstalk via promiscuous ligand-receptor interactions (Nakae et al. 2001). Intracellular downstream signaling cascades are nevertheless highly shared between IGF and insulin receptors, and current efforts are trying to define how signaling specificity is encoded in the transduction between cell membrane and the nucleus (Tseng et al. 2002; 2004). With the separation into IGF and insulin pathways at the cell surface, a group of six high affinity IGF-binding proteins (IGFBP) made their appearance (Hwa et al. 1999). Produced by many different cell types from embryonic development onward, but also throughout adult life, these IGFBPs are present in the circulation and are also secreted into the interstitial spaces. Their main action seems to be the specific reduction of IGF availability to the receptor, in other words an inhibition of IGF signaling to its cognate receptor, although most of the in vivo function of IGFBPs was established through transgenic models using IGF binding protein overexpression (Dupont and Holzenberger 2003). By contrast, gene knockout models, including combinations of several IGFBP inactivations, could not reveal phenotypic traits nearly as informative as the respective knockouts of the IGF receptor or of the IGF-1 or IGF-II ligands. The IGF receptor knockout is lethal at birth, due to severe growth retardation and immature respiratory function, whereas the IGF-1 knockout engenders severe intrauterine and profound postnatal growth retardation (Baker et al. 1993; Liu et al. 1993). Besides somatic growth, IGF signaling is capable of regulating energy storage through the control of fat storage and, importantly, is a major regulator of life span, probably via coordinated control of a number of genes that are responsible for oxidative stress responses in the cell (Murphy et al. 2003).

IGF signaling is highly pleiotropic. It is implicated in tissue regeneration and vascularization, including neo-vascularization (Kondo et al. 2003), and facilitates liver repair (manuscript in preparation). It also plays a major role in tumor growth, although some tumors apparently do not depend on IGF-1R, as we recently showed using a transgenic hepatocarcinoma mouse model (Cadoret et al. 2005). IGF signaling also participates significantly in the maintenance of glucose homeostasis, as was demonstrated recently in mice (Kulkarni et al. 2002).

Somatotrope control of IGF and growth

Somatic growth regulation has a strong central component. During postnatal growth, a set of highly specialized neurons in the mammalian hypothalamus secretes GHRH, a peptide directed by axonal transport and subtle secretory mechanisms via the venous blood stream to the somatotrope cell in the pituitary gland. Activation of GHRH receptors on the pituitary somatotrope cells then stimulates GH production. This peptide hormone is subject to a particular pulsatile secretion into the general circulation. The pattern of this secretion is highly sex-dimorphic and is thought to be crucially involved in the sex dimorphism of mammalian growth. In the periphery, circulating GH has direct effects and indirect effects, the latter being defined as depending on the stimulation of local IGF production and secretion. The liver IGF-I production is particularly sensitive to GH, and most of the circulating IGF-I, which is also almost completely bound to IGFBP-3 and ALS , is actually synthesized by the liver (Sjogren et al. 1999; Yakar et al. 1999).

This hypothalamic-pituitary hormonal axis efficiently controls peripheral growth but, depending on which genes are inactivated in this axis, the resulting phenotypes in the mouse show considerable differences, especially beyond growth characteristics. In Snell dwarf and Ames dwarf mutants, pituitary differentiation is profoundly disturbed, leading invariably to a complete lack of peripheral GH, TSH, LH and FSH (Brown-Borg et al. 1996; Flurkey et al. 2001). Although interpretation of the resulting phenotypes is somehow difficult due to this constitutive panhypopituitarism, the growth deficit phenotype of these mutants indicates that a large part of their postnatal growth is controlled by GH. Snell and Ames dwarfs, which are homozygous null mutants for Pit-1 and Prop1, respectively, are sterile, probably because of the lack of LH and FSH. They also lack TSH, a deficiency that could contribute to both the growth and fertility phenotypes. GHRH receptor (GHRHR) knockout mice (Little mice) and GHR/BP knockout mice (Coschigano et al. 2000) are also small and subfertile, underscoring the fact that somatotrope signals and GH signals play a definitive role in the development and maintenance of fertility and reproductive function. IGF-1R knockdown mutations (Holzenberger et al. 2000b; 2001) mostly affect postnatal growth and the growth and differentiation of specific tissues like the adipose tissue, but without affecting fertility, at least in the ranges of inactivation that we were studying. Finally, mutations downstream of IGF-1R, like the targeted inactivation of the p66 Shc isoforms, do not affect growth but do specifically alter the susceptibility of the animal to oxidative stresses (Migliaccio et al. 1999). Together, these findings argue in favor of a neuroendocrine-endocrine signaling network that efficiently integrates several vital functions (growth, fertility/reproduction, and longevity) and is capable of co-regulating several of them, or one by one, depending on the level on which the gene of interest is acting and also the degree of gene inactivation. However, the fine-tuning of this regulation is still incompletely understood.

Using homologous recombination in the mouse, we have created over the last few years a series of mutant IGF-1R alleles (Holzenberger et al. 2000a; 2000b). Through a combination of these alleles, we produced additional compound heterozygous mutants with vastly different levels of IGF-1R (Holzenberger et al. 2001). The relationship that exists between IGF-1R levels and body size in young adult mice was determined from data extracted from several of our publications and is shown in Figure 2.

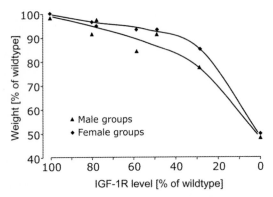

Fig. 2. IGF-1R levels and adult growth deficits in mice. Data were collected from seven independent experiments where postnatal growth was measured in cohorts of mice with different degrees of IGF-1R inactivations (see Fig. 3 for details). Mice in the last group correspond to the classical IGF-1R knockout and have no receptor. These mutants were not viable and therefore this measurement was at birth, not in adults. There seems to be a roughly linear relationship between IGF-1R prevalence and growth in mice between 30 and 100% of normal receptor levels. The slope for males is approximately twice the slope for females, indicating that the postnatal growth in males depends more on IGF-1R than it does in females.

IGF-1R mutations in mice

The first of these mutants was a hypomorphic IGF-1R allele obtained through introduction of a neomycine selection cassette into intron 2 of the IGF-1R gene (Holzenberger et al. 2000b). Through aberrant splicing, this intronic mutation reduces the expression levels of IGF receptor by 40%, as measured by IGF ligand binding assay. Thus, in the heterozygous state, IGF-1R levels are reduced to 80% of the wild type and, in the homozygous state, to around 60%. Surprisingly, these mutants show a normal growth pattern until three weeks after birth. After this, they progressively develop slight to moderate growth deficits. The plot of their respective growth velocities indicates that the peak values prior to the age of onset of fertility are reduced in these mutants compared to their wild type littermates (Fig. 3A and B). Using a gene dosage approach, we then created mice with up to 80% inactivation of IGF-1R (Holzenberger et al. 2001). While these mice were considerably smaller as adults compared to their wild type littermates, the onset of this grow deficit was again around two to three weeks after birth (Fig. 3C). In a third experiment, we used heterozygous IGF-1R knockout mutants. Results obtained with that group of mutants were, in principle, very similar to the two previous IGF-1R mutants, since the onset of the growth deficit occurred again at three weeks postnatally and also because mutant males were slightly more affected than mutant females (Holzenberger et al. 2003). However, and in addition to this, our data revealed that the IGF-1R-dependent growth velocity peak actually preceded the sex-related growth velocity peak (Fig. 3D and 3E), and both peaks were nearly one week apart (Holzenberger 2004). Together, these results suggested that partial IGF-1R insufficiencies of different degree produce defects

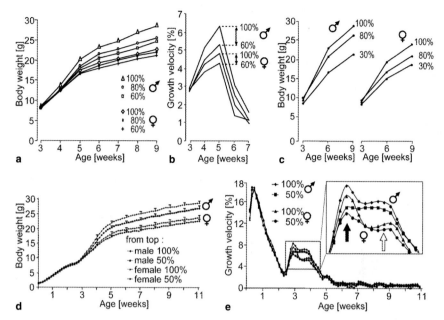

Fig. 3. Postnatal growth and growth velocity in different mouse models of genetic IGF-1R inactivation. **A.** IGF-1R knock-down alleles produce a postnatal growth deficit. Growth curves for male and female animals with either 60% or 80% of normal receptor levels were compared to wild type control littermates (100% IGF-1R) (from Holzenberger et al. 2000b). **B.** Growth velocity (weight gain per 24 hours expressed in percent of absolute body weight) from the data of the 60% and 100% groups in A. Double arrows indicate the reduction in "prepubertal" peak velocity. **C.** IGF-1R inactivation by gene dosage (from Holzenberger et al. 2000b). Receptor deficiencies down to 30% were obtained and differences in mean body weight observed at 3, 6, and 9 weeks of age. **D.** Mice with 50% of wild type receptor levels (IGF-1R⁺/⁻ mutants) and wild type siblings showed identical growth until day 20. Thereafter, during the growth spurt (weeks 3 to 5), slight deficits in weight gain appeared (from Holzenberger et al. 2003). **E.** In these IGF-1R⁺/⁻ mutants, the growth velocity peaks at three weeks of age (black arrow, magnification in inset) are blunted, whereas the sex-dependent peaks (white arrow) are not. Curves in **D** and **E** were established using a sliding mean over three consecutive means (from Holzenberger 2004).

in growth velocity during a relatively short time window of postnatal development that was situated around the time point of normal weaning and shortly before male and female mice become fertile. It appears that the partially IGF-1R-deficient mice change their individual growth trajectories when that first growth velocity peak occurs. Although these altered trajectories appear as a continuous defect after the age of three weeks, we propose to interpret this altered growth as the long-term consequence of a more limited initial developmental defect. In fact, many classical knockout or knockdown phenotypes occur in this manner, often as cumulative disorders of discrete defects, with this being rather the rule than the exception. Due to the precise onset of the deficit, we believe that it was of central, neurendocrine origin rather than an extended peripheral defect.

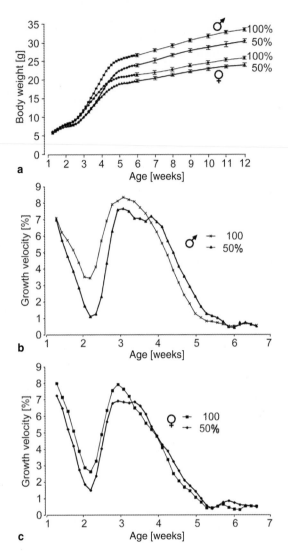

Fig. 4. Growth in mice that have a brain-specific heterozygous IGF-1R knockout, resulting in 50% of normal IGF-1R levels in the brain. A. Postnatal weight was, from day 18 onwards, significantly lower in 50% of males and females compared with wild type controls. Error bars indicate SEM. B. Growth velocity in males with 50% IGF-1R levels in the brain is shifted to lower values during postnatal weeks 2-4. The peak at three weeks is blunted. C. Similar changes as in B are observed in females.

To find out more about the localization of this hypothetical developmental defect, we produced conditional heterozygous IGF-1R knockout models using the Cre-lox system (manuscript in preparation). These mice lack one functional IGF-1R allele specifically in the brain. Therefore, their IGF-1R levels are reduced to half in the CNS, whereas peripheral receptor levels remain completely normal. These mutant animals develop a severe secretory defect for GH very early during postnatal life and, secondary to that, a significant lack of circulating IGF-I (see forthcoming publication for details). As a result of this somatotrope deprivation of neuorendocrine origin, these mice develop a postnatal growth deficit with very similar characteristics as the above-described phenotypes in IGF-1R knockout or knock-down models. In these mice, the sex-dimorphic growth deficit is due to the particularly low peripheral IGF-1 levels in the males. As in the previous models, the growth deficit is definitive and mice do not show catch-up growth during early adulthood (Fig. 4A). Additional insight comes again from the corresponding growth velocity curves (Fig. 4B and 4C). The genetic defect alters the early postnatal growth pattern. Peak growth velocity at around three weeks of age is diminished in male and female mutant mice. This defect definitively changes the weight gain trajectory of the mutant animals. Since this conditional mutation does not extend into the pituitary gland (due to the fact that the anterior pituitary is not of neuroepithelial origin), we deduce that the cellular mechanisms that control the "prepubertal" growth spurt in mice is located within the central nervous system, and we suggest that the pathophysiology of hypothalamic GHRH neurons may play a key role in this type of growth spurt regulation.

Overview

Somatic growth is regulated by conserved mechanisms and pathways in animals. The pathophysiology of the somatotrope function in mammals in particular has been studied using genetically modified models. Results from the last decade have shown that growth is a very plastic process, controlled by a large set of genes. The IGF and growth hormone gene family plays a key role in this growth regulation. We produced several mouse models of IGF-1R insufficiency by homologous recombination. We showed that partial inactivation of the IGF-1R can produce postnatal growth deficits. These defects appear during the postnatal growth spurt and persist into adulthood. Growth deficits due to IGF-1R inactivation vary with the dosage of the IGF-1R gene. It appears that the postnatal growth of males relies more strongly on IGF-1R levels than the growth of females. IGF-1R in the brain could play a particularly important role in the development of the somatotrope function in mammals. Other vital functions, like the cellular resistance to oxidative stress, are also regulated via the somatotrope axis. Although a link between GH/IGF signaling and longevity has been found in various mouse models, it is at present unclear whether this finding applies to humans, too.

References

Abuzzahab MJ, Schneider A, Goddard A, Grigorescu F, Lautier C, Keller E, Kiess W, Klammt J, Kratzsch J, Osgood D, Pfäffle R, Raile K, Seidel B, Smith RJ, Chernausek SD (2003) IGF-I receptor mutations resulting in intrauterine and postnatal growth retardation. New Engl J Med 349: 2211-2222

Baker J, Liu J-P, Robertson EJ, Efstratiadis A (1993) Role of insulin-like growth factors in embryonic and postnatal growth. Cell 75: 73–82

Brown-Borg HM, Borg KE, Meliska CJ, Bartke A (1996) Dwarf mice and the ageing process. Nature 384: 33

Cadoret A, Desbois-Mouthon C, Wendum D, Leneuve P, Perret C, Tronche F, Housset C, Holzenberger M (2005) c-Myc-induced hepatocarcinogenesis in the absence of insulin-like growth factor 1 receptor. Int J Cancer 114: 668-672

Coschigano KT, Clemmons D, Bellushi LL, Kopchick JJ (2000) Assessment of growth parameters and life span of GHR/BP gene-disrupted mice. Endocrinology 141: 2608-2613

Denley A, Wang CC, McNeil KA, Walenkamp MJE, van Duyvenvoorde H, Wit JM, Wallace JC, Norton RS, Karperien M, Forbes BE (2005) Structural and functional characteristics of the Val44Met IGF-I missense mutation: correlation with effects on growth and development. Mol Endocrinol 19: 711-721

Dupont J, Holzenberger M (2003) Biology of insulin-like growth factors in development. Birth Defects Res (Part C) 69: 257-271

Flurkey K, Papaconstantinou J, Miller RA, Harrison DE (2001) Lifespan extension and delayed immune and collagen aging in mutant mice with defects in growth hormone production. Proc Natl Acad Sci USA 9: 6736–6741

Gems D, Partridge L (2001) Insulin/IGF signalling and ageing: seeing the bigger picture. Curr Opin Genet Dev 11: 287–292

Guarente L, Kenyon C (2000) Genetic pathways that regulate ageing in model organisms. Nature 408: 255–262

Holzenberger M (2004) The GH/IGF-1 axis and longevity. Eur J Endocrinol 151, suppl 1: S23-27

Holzenberger M, Martín-Crespo RM, Vicent D, Ruiz-Torres A (1991) Decelerated growth and longevity in men. Arch Gerontol Geriat 13: 89-101

Holzenberger M, Lenzner C, Leneuve P, Zaoui R, Hamard G, Vaulont S, Le Bouc Y (2000a) Cre-mediated germ-line mosaicism: a method allowing rapid generation of several alleles of a target gene. Nucleic Acids Res 28: e92

Holzenberger M, Leneuve P, Hamard G, Ducos B, Périn L, Binoux M, Le Bouc Y (2000b) A targeted partial invalidation of the IGF-I receptor gene in mice causes a postnatal growth deficit. Endocrinology 141: 2557-2566

Holzenberger M, Hamard G, Zaoui R, Leneuve P, Ducos B, Beccavin C, Périn L, Le Bouc Y (2001) IGF-I receptor gene dosage generates a sexually dimorphic pattern of organ-specific growth deficits, affecting fat tissue in particular. Endocrinology 142: 4469-4478

Holzenberger M, Dupont J, Ducos B, Leneuve P, Géloën A, Even PC, Cervera P, Le Bouc Y (2003) IGF-1 receptor regulates life span and resistance to oxidative stress in mice. Nature 421: 182-187

Hwa V, Oh Y, Burren CP, Choi WK, Graham DL, Ingermann A, Kim H-S, Lopez-Bermejo A, Minniti G, Nagalla SR, Pai K, Spagnoli A, Vorwerk P, Wanek DLV, Wilson EM, Yamanaka Y, Yang DH, Rosenfeld RG (1999) The IGF binding protein superfamily. In: Rosenfeld RG, Roberts CT (eds) The IGF system. Humana Press, Totowa, NJ, pp. 315-328

Kenyon C (2001) A conserved regulatory system for aging. Cell 105: 165–168

Kondo T, Vicent D, Suzuma K, Yanagisawa M, King GL, Holzenberger M, Kahn CR (2003) Knockout of insulin and IGF-1 receptors on vascular endothelial cells protects against retinal neovascularization. J Clin Invest 111: 1835-1842

Kulkarni RN, Holzenberger M, Shih DQ, Ozcan U, Stoffel M, Magnuson MA, Kahn CR (2002) ß-cell-specific deletion of the Igf-1 receptor leads to hyperinsulinemia and glucose intolerance but does not alter β-cell mass. Nature Genet 31: 111-115

Liu JP, Baker J, Perkins AS, Robertson EJ, Efstratiadis A (1993) Mice carrying null mutations of the genes encoding insulin-like growth factor I (Igf-1) and type 1 IGF receptor (Igf1r). Cell 75: 59–72

Lupu L, Terwilliger JD, Lee K, Segre GV, Efstratiadis A (2001) Roles of growth hormone and insulin-like growth factor 1 in mouse postnatal growth. Dev Biol 229, 141–162

Migliaccio E, Giorgio M, Mele S, Pelicci G, Reboldi P, Pandolfi PP, Lanfrancone L, Pelicci PG. (1999) The p66shc adaptor protein controls oxidative stress response and life span in mammals. Nature 402: 309–313

Murakami S, Salmon A, Miller RA (2003) Multiplex stress resistance in cells from long-lived dwarf mice. FASEB J 17: 1565-1566

Murphy CT, McCarroll SA, Bargmann CI, Fraser A, Kamath RS, Ahringer J, Li H, Kenyon C (2003) Genes that act downstream of DAF-16 to influence the lifespan of Caenorhabditis elegans. Nature 424: 277-283

Nakae J, Kido Y, Accili D (2001) Distinct and overlapping functions of insulin and IGF-I receptors. Endocr Rev 22: 818-835

Sjogren K, Liu JL, Blad K, Skrtic S, Vidal O, Wallenius V, LeRoith D, Tornell J, Isaksson OG, Jansson JO, Ohlsson C (1999) Liver-derived insulin-like growth factor I (IGF-I) is the principal source of IGF-I in blood but is not required for postnatal body growth in mice. Proc Natl Acad Sci USA 96: 7088-7092

Tseng YH, Ueki K, Kriauciunas KM, Kahn CR (2002) Differential roles of insulin receptor substrates in the anti-apoptotic function of insulin-like growth factor-1 and insulin. J Biol Chem 277: 31601-31611

Tseng YH, Kriauciunas KM, Kokkotou E, Kahn CR (2004) Differential roles of insulin receptor substrates in brown adipocyte differentiation. Mol Cell Biol 24: 1918-1929

Ullrich A, Bell JR, Chen EY, Herrera R, Petruzelli LM, Dull TJ, Gray A, Coussens L, Liao YC, Tsubokawa M, Mason A, Seeburg PH, Grunfeld C, Rosen OM, Ramachandran J (1985) Human insulin receptor and its relationship to the tyrosine kinase family of oncogenes. Nature 313: 756–761

Ullrich A, Gray A, Tam AW, Yang-Feng T, Tsubokawa M, Collins C, Henzel W, Le Bon T, Kathuria S, Chen E, Jacobs S, Francke U, Ramachandran J, Fujita-Yamaguchi Y (1986) Insulin-like growth factor-I receptor primary structure: Comparison with insulin receptor suggests structural determinants that define functional specificity. EMBO J 5: 2503–2512

Woods KA, Camacho-Hübner C, Savage MO, Clark AJ (1996) Intrauterine growth retardation and postnatal growth failure associated with deletion of the insulin-like growth factor I gene. New Engl J Med 335: 1363-1367

Yakar S, Liu JL, Stannard B, Butler A, Accili D, Sauer B, LeRoith D (1999) Normal growth and development in the absence of hepatic insulin-like growth factor I. Proc Natl Acad Sci USA 96:7324-7329

Downstream Mechanisms of Growth Hormone Action

Vivian Hwa[1] and Ron G. Rosenfeld[1,2,3]

Summary

Growth hormone (GH) activates a number of signaling pathways upon binding to its cognate receptor (GHR). Insights into downstream mechanisms of GH actions have been gained through the recent identification of a homozygous *STAT5b* gene mutation in a young patient presenting with severe growth failure (height −7.5 SD) associated with normal GHR, elevated serum concentrations of GH, and markedly reduced serum IGF-I. At the cellular level, GH-induced genes that are STAT5b-dependent, such as IGF-I, were dysregulated, whereas regulation of other GH-responsive genes, such as SOCS2 and SOCS3, was unimpaired. The implication is that STAT5b has a unique and critical role in the growth-mediating actions of GH through regulating IGF-I expression. This finding is supported by the identification of a second case of *STAT5b* mutation associated with GH insensitivity. The role(s) of the other signaling pathways has yet to be fully characterized. Clearly, the unmasking of the molecular bases for cases of GH insensitivity will greatly increase our understanding of both normal and aberrant human growth.

Introduction

The growth-promoting effects of growth hormone (GH) are mediated primarily through regulating expression of insulin-like growth factor–I (IGF-I), both circulating and peripheral, as demonstrated in rodent models and in case studies in humans. GH, upon binding to its cognate receptor, the GH receptor (GHR), initiates signaling by activation of receptor-associated Janus kinase 2 (JAK2), which undergoes autophosphorylation and, concurrently, phosphorylates tyrosines on GHR. The phospho-tyrosines on GHR serve as docking sites for components of at least three pathways: the STAT (signal transducer and activators of transcription), the MAPK (mitogen-activated protein kinase), and the PI3K (phosphoinositide-3 kinase) pathways (Fig. 1). The signaling cascade culminates

[1] Department of Pediatrics, Oregon Health and Sciences University, Portland , OR, USA
[2] Lucile Packard Foundation for Children's Health, Palo Alto, California, USA
[3] Department of Pediatrics, Stanford University, Stanford, California, USA

Carel et al.
Deciphering Growth
© Springer-Verlag Berlin Heidelberg

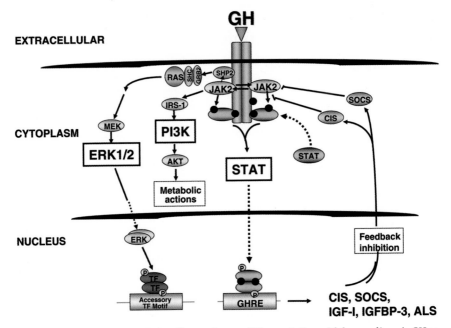

Fig. 1. Growth hormone (GH) signaling pathways. GH association with homo-dimeric GH receptor results in recruitment of JAK2 and subsequent activation of the MAPK-ERK1/2, PI3K, and STAT pathways.

Abbreviations: SHP, SH2-containing phosphatase; GRB, growth factor receptor-bound protein; SHC, SH2-containing collagen-related protein; RAS, small GTP binding protein; MEK, mitogen-activated protein kinase/ERK kinase; ERK, extracellular signal-regulated kinase; IRS, insulin receptor substrate; PI3K, phosphoinositide 3 kinase; AKT, AKT8 virus oncogene cellular homolog; JAK, Janus-family tyrosine kinase; STAT, signal transducer and activator of transcription; CIS, cytokine inducible SH2-containing protein; SOCS, suppressor of cytokine signaling; TF, transcription factor, GHRE, growth hormone response element; IGF-I, insulin-like growth factor-1; IGFBP-3, IGF binding protein 3; ALS, acid labile subunit.

in the regulation of multiple genes, including IGF-I and IGF binding protein-3 (IGFBP-3). Defects in any of these pathways that result in dysregulation of IGF-I expression would be predicted to impact normal growth.

The deciphering of GH-mediated regulation of IGF-I expression has depended largely on studies employing rodent models and reconstitution systems, with limited supporting data in humans. Our current understanding of the GH-IGF-I axis in humans has come from case studies of GH insensitivity (GHI), a phenotype clinically indistinguishable from severe congenital GH deficiency (GHD), with minimal growth retardation in utero, infantile facial appearance, profound postnatal growth failure (Rosenfeld et al. 1994), and markedly reduced serum concentrations of IGF-I and IGFBP-3 (Buckway et al. 2001; Dattani and Preece 2004; Savage et al. 2001). The clinical difference is the demonstration of resistance to exogenous growth hormone therapy in patients with GHI (Savage et al. 2001; Savage and Rosenfeld 1999).

The molecular basis for GHI has been limited to identification of defects in the GH receptor (GHR), first described by Laron et al. (1966), and one case report of deletion of the *IGF-I* gene (Woods et al., 1996), but the majority of GHI cases remained largely uninvestigated. In particular, GHI with normal GHR has been inadequately studied, yet these cases are likely to yield the most information on mechanism(s) of downstream GH action, as they are most probably consequences of defective GHR signaling. It is notable that, since many components of the GH-GHR signaling pathways are also activated by other cytokines and growth factors, defects in these intracellular components may manifest in clinical phenotypes in addition to resistance to GH. Hence, a complex clinical phenotype that includes GHI with normal GHR would be strongly indicative of post-GHR defects.

In this report, we summarize the role of the three main GHR signaling pathways in GH-induced regulation of IGF-I expression and in growth. Insights gained from the first report of a GHR signaling defect, a homozygous missense STAT5b mutation associated with GHI (Kofoed et al., 2003), will be emphasized.

JAK2 is essential for initiating GHR signaling

Since the GHR lacks intrinsic kinase activity, the recruitment of cytosolic kinase JAK2 upon GH-GHR association is crucial in the initiation of the GHR signaling cascades. This has been demonstrated by studies showing that the loss of ability of GHR to bind JAK2 (due to mutations in sequences designated Box 1; Yi et al. 1996) results in loss of GH-induced GHR signaling, and by a recent study that demonstrated that conditional knockout of the JAK2 gene in mice decoupled GHR from down-stream signaling (Krempler et al. 2004). JAK2, like the other three members of the JAK family (Jak1, JAK3 and Tyk), can associate with multiple ligand-activated receptors (Leonard and O'Shea 1998; Levy and Darnell 2002). The homodimeric GHR, unlike other cytokine receptors, does not appear to associate with members of the JAK family, other than JAK2. To date, no human mutations in JAK2 have been identified. Homozygous mutation in JAK2 in humans may be lethal, as mouse models carrying targeted disruption of the *JAK2* gene displayed embryonic lethality due to failure of erythropoiesis (Krempler et al. 2004; Neubauer et al. 1998; Parganas et al. 1998).

MAPK-ERK1/2 pathway in GHR signaling

The role of the MAPK-ERK1/2 signaling pathway in mediating GH action is not well defined. One of the well-established functions of activated ERK1/2 is phosphorylation of pertinent serine residues on transcription factors, thereby modulating transcriptional activity. This function has been demonstrated for a number of STATs, specifically STAT1, -3 and –4 (Levy and Darnell 2002). Since STAT5a/b is important in GH-induced signaling (see below), the role of the ERK1/2 pathway in modulating STAT5 activity has been investigated. The studies have shown that, although STAT5a/b is serine phosphorylated upon GH

(Park et al. 2001; Pircher et al. 1997; Shoba et al. 2001), prolactin (Yamashita et al. 1998), or interleukin2 (Nagy et al. 2002) stimulation, serine phosphorylation does not appear to be mediated by pERK1/2, as the phosphorylation process was insensitive to the MAPK inhibitor, PD98059. More importantly, GH induction of IGF-I expression in primary rat hepatocytes was unaffected by PD98059, suggesting that the MAPK-ERK1/2 pathway does not play a critical role in regulating IGF-I expression. Consistent with this finding was the observation that targeted disruption of ERK1 (p44) in rodent models resulted in mice that were viable, fertile, and of normal size, but were defective in thymocyte maturation (Pages et al. 1999), and targeted disruption of ERK2 (p42) was embryonically lethal (Saba-El-Leil et al. 2003). Similar mutations in humans have yet to be identified. Altogether, the data do not support the involvement of GH-activated MAPK-ERK1/2 pathway in the regulation of IGF-I expression.

Role of PI3K-AKT pathway in GH action

The PI3K-AKT pathway, activated by multiple growth factors and cytokines, is well documented to promote cell proliferation and differentiation and to be anti-apoptotic. The role of the PI3K-AKT pathway in GH action is less well understood, although blocking of this pathway with the PI3K pharmaceutical inhibitor, LY294002, resulted in reduction of GH-induced regulation of IGF-I expression in mouse cells (Frost et al. 2002; Shoba et al. 2001). This finding suggested that the PI3K pathway may participate in GH-induced regulation of IGF-I expression (Frost et al. 2002; Shoba et al. 2001). Interestingly, rodent studies indicated that ablation of the p85α regulatory subunit of PI3K by targeted gene disruption did not impede growth of the mice. However, p85α$^{-/-}$ mice exhibited Xid-like immunodeficiency and succumbed to bacterial infection if not kept in a pathogen-free environment (Fruman et al. 1999; Suzuki et al. 1999).

Targeted gene disruption of the downstream effector of PI3K, AKT, of which there are three isoforms, has also been investigated. Strikingly, the double knockout AKT1$^{-/-}$2$^{-/-}$ mouse showed a phenotype similar to the IGF-I receptor (IGF-IR)$^{-/-}$ mouse (Liu et al. 1993), that is, dwarfism, impaired skin and bone development, impeded adipogenesis, skeletal muscle atrophy, and death shortly after birth (Peng et al. 2003). These data indicate that the PI3K-AKT pathway is essential for survival and normal growth, although not necessarily as a direct function of GH action.

Critical role of STAT5b in mediating GH action

STAT proteins are unique cytosolic proteins that function both as signal transducers and, upon activation, as transcription factors. Discovered over 10 years ago, there are seven known mammalian STATs that participate in a plethora of biological activities (Levy and Darnell 2002). Structurally similar (Fig. 2), all the STATs can be activated after one or more cytokines and/or growth factors interact with their cognate receptors. STAT proteins dock, via their src-homology-2

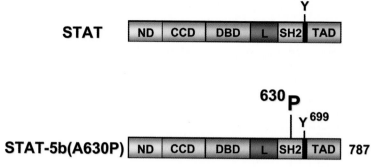

Fig. 2. Schematic structure of human STAT proteins. STATs are composed of six protein modules: ND, amino-terminal domain; CCD, coiled-coil domain; DBD, DNA-binding domain; L, linker domain' SH2, *src*-homology2 domain; and TAD, transactivation domain. The tyrosine (Y) that is phosphorylated is indicated. The missense mutation in *STAT5b* gene, identified in GHI subject #1, results in Alanine to Proline substitution at position 630.

(SH2) domain, to phospho-tyrosines on ligand-activated receptors. The docked STATs are subsequently phosphorylated on single tyrosines at the C-terminus of the protein by receptor-associated JAKs, and then dimerize and translocate to the nucleus, where they bind to DNA through their DNA binding domain (DBD). Serine phosphorylation and interactions with other transcription factors modulate the efficiency of STAT transcription.

Reconstitution studies and rodent models have demonstrated that GH activates STAT1, 3, and isoforms 5a and 5b. Of these four STAT proteins, there has been a steady accumulation of data supporting the direct involvement of STAT5b in regulating IGF-I expression, including the recent identification of STAT5b response elements in intron 2 of the rat *IGF-I* gene (Woelfle et al. 2003). Gene disruption studies in rodent models, in particular, implicated STAT5b as critical for growth. STAT5b$^{-/-}$ mice displayed loss of sexual dimorphic growth, with concomitant reduction in circulating levels of IGF-I (Teglund et al. 1998; Udy et al. 1997). Male STAT5b$^{-/-}$ mice were reduced to the size of female mice, but no size differences were observed between female STAT5b$^{-/-}$ mice and wild-type females. In both male and female STAT5b$^{-/-}$ mice, circulating levels of IGF-I were reduced to 50-70% of wild-type levels. Thus, the biological effects of STAT5b on regulation of circulating IGF-I and on growth in mice, although significant, were, on the whole, relatively modest.

In humans, the recent identification of the first case of a *STAT5b* mutation associated with GHI (Kofoed et al. 2003) has provided insights into downstream mechanisms of GH action. The only other human *STAT* mutations are those identified in the *STAT1* gene, where the associated phenotype was characterized by increased susceptibility to *Mycobacterium* infections (Dupuis et al. 2001, 2003). The *STAT5b* mutation, a homozygous missense mutation in Exon 15 of the *STAT5b* gene, was identified in a young female subject who presented a complex phenotype of GHI and symptoms consistent with immune dysfunction (Kofoed et al. 2003; Fig. 2). Diagnosis of GHI was based, in part, on the unaltered growth

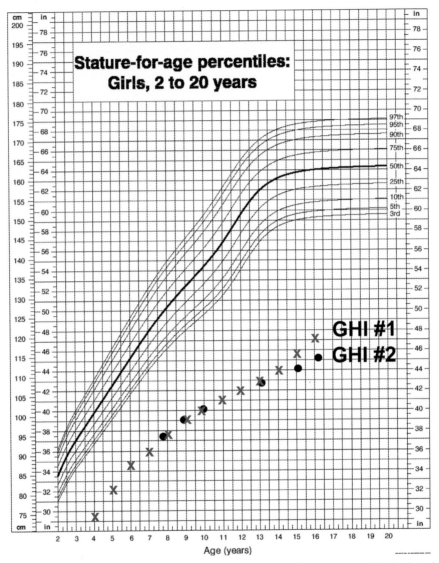

Fig. 3. Growth profiles of GHI subjects. GHI #1 has been previously reported (Kofoed et al. 2003). GHI #2 demonstrates an identical growth profile to subject #1.

velocity of the subject after one year of exogenous GH therapy and was confirmed by a failed IGF-I generation test; the subsequent demonstration of a normal GHR gene suggested a potential post-GHR defect (Kofoed et al. 2003).

The subject was the product of a consanguineous marriage, and although birth weight and length were normal, her post-natal growth curve was similar to those of classical GHI patients (Fig. 3). At age 16.5 years, her height was 117.8 cm (-7.5

SDS), with normal body proportions and delayed puberty. Her circulating GH concentrations were normal-elevated (baseline GH: 9.4 ng/ml; stimulated GH > 50 ng/ml), but serum IGF-I was abnormally low (<15% of normal for age), as was IGFBP-3 (<35% of normal).

The correlation of a *STAT5b* mutation with the low circulating IGF-I levels and severe growth retardation has been supported by the recent identification of a second case of GHI associated with a *STAT5b* mutation (unpublished).

In addition to insights into the clinical phenotype, study of the first mutant STAT5b protein, itself, has provided information on STAT5b function. The missense mutation in *STAT5b* gene generated a protein in which Alanine at residue 630 was substituted for Proline (A630P). A630 resides within the conserved SH2 domain (Fig. 2) of the protein, a domain essential for docking of the STAT5b to phospho-tyrosines on ligand-activated receptors, for homo-dimerization and for subsequent stabilization of STAT-DNA complexes. The consequences of the A630P mutation in STAT5b included poor detection of the mutant protein on immunoblot analysis and, in both primary dermal fibroblasts and in reconstitution systems, GH could not induce phosphorylation of mutant STAT5b(A630P) and STAT5b(A630P) could not drive gene expression (neither IGF-I nor a luciferase-reporter construct; Fig. 4). These results are consistent with the inability of STAT5b(A630P) to dock to GHR prior to a phosphorylation event. Other cytokines, such as interferon-gamma (IFN-γ), could not activate the mutant STAT5b either (Hwa et al. 2004), thereby supporting the hypothesis that a number of ligand-mediated receptor signaling systems were aberrant, which could explain the complex phenotype of GHI and immune dysfunction exhibited by the subject.

Primary cells lacking functional STAT5b proteins may provide unique insights into genes regulated by GH. While GH-induced regulation of IGF-I and IGFBP-3 is clearly STAT5b-dependent, not all genes regulated by GH-GHR signaling pathways are dependent on a functional STAT5b. For example, GH-induced regulation of SOCS2 and SOCS3, two proteins involved in the negative feed-back loop of the JAK-STAT pathway (Cooney 2002; Ram and Waxman 1999; Fig. 1), was similar to that observed in normal fibroblasts (Fig. 5). Further investigations should reveal the GH-induced signaling pathway responsible for this regulation and the consequent biological effects. Altogether, the implication is that STAT5b has a novel and critical role in the growth–mediating actions of GH through regulating IGF-I expression.

Conclusions

It has only been in recent years that the range in molecular defects that can contribute to growth disorders has become fully evident. Insights into downstream GH actions in humans have been gained through biochemical and molecular analysis of cases of GH insensitivity with normal GHR, as was demonstrated with the identification of the first patient with GHI associated with a *STAT5b* mutation. Clearly, continued unmasking of the molecular basis for GHI will contribute greatly to our future understanding of the GH-IGF axis in normal and aberrant human growth.

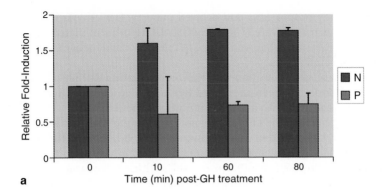

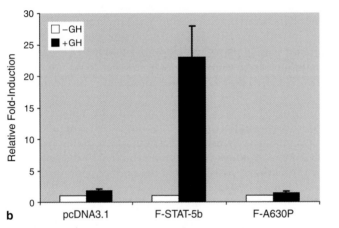

Fig. 4. STAT5b(A630P) demonstrates a lack of transcriptional activity. A. Primary dermal fibroblasts were incubated with GH (500 ng/ml) for the times indicated. Total RNA was collected, and the response of IGF-I mRNA to GH was analyzed by real-time quantitative polymerase chain reaction. Results are normalized to those for 18S and are expressed as the relative-fold induction compared with untreated cells. Data are the means (±SD) from two independent experiments done in triplicate. (Kofoed et al. 2003). B. N-terminally FLAG-tagged recombinant wild-type (F-STAT5b) and mutant STAT5b(A630P) (F-A630P) were generated and employed in reconstitution experiments in HEK293 cells stably overexpressing human GHR (courtesy of Dr R. Ross, Sheffield, England). Cells were co-transfected with GHRE(Spi2.1)-luciferase reporter construct and plasmids pcDNA3.1 (vector), F-STAT5b or F-A630P, and treated with GH. After 24 h, cell lysates were collected for determination of luciferase activity. Results are reported as relative fold-induction compared to untreated (-GH, given an arbitrary value of 1) ± SD, from four independent experiments, each performed in triplicate. N. Normal, P. Patient.

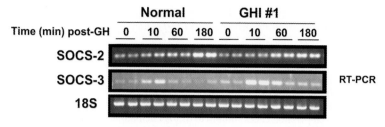

Fig. 5. GH-induced, STAT5b-independent, regulation of genes. Primary dermal fibroblasts were incubated with GH (500 ng/ml) for the times indicated. Total RNA was collected, and the response of SOCS2 and SOCS3 mRNA to GH was analyzed by reverse-transcriptase-polymerase chain reaction (RT-PCR). RT-PCR of 18S was employed as control.

References

Buckway CK, Guevara-Aguirre J, Pratt KL, Burren CP, Rosenfeld RG (2001) The IGF-I generation test revisited: a marker of GH sensitivity. J Clin Endocrinol Metab 86: 5176-5183.

Cooney RN (2002) Suppressors of cytokine signaling (SOCS): inhibitors of the JAK/STAT pathway. Shock 17: 83-90.

Dattani M,Preece MA (2004) Growth hormone deficiency and related disorders: insights into causation, diagnosis, and treatment. Lancet 363: 1977-1987.

Dupuis S, Dargemont C, Fieschi C, Thomassin N, Rosenzweig S, Harris J, Holland SM, Schreiber RD, Casanova J-L (2001) Impairment of Mycobacteril but not viral immunity by a germline human STAT1 mutation. Science 293: 300-303.

Dupuis S, Jouanguy E, Al-Hajjar S, Fieschi C., Al-Mohsen I.Z, Al-Jumaah S, Yang K., Chapgier A, Eidenschenk C., Eid P, Al Ghonaium A, Tufenkeji H, Frayha H, Al-Gazlan S, Al-Rayes H, Schreiber RD, Gresser I, Casanova JL (2003) Impaired response to interferon-a/ß and lethal viral disease in human STAT1 deficiency. Nature Genet 33: 388-391.

Frost RA, Nystrom GJ, Lang CH (2002) Regulation of IGF-I mRNA and signal transducers and activators of transcription-3 and -5 (Stat-3 and -5) by GH in C2C12 myoblasts. Endocrinology 143: 492-503.

Fruman DA, Snapper SB, Yballe CM, Davidson L, Yu JY, Alt FW, Cantley LC (1999) Impaired B cell development and proliferation in absence of phosphoinositide 3-kinase p85alpha. Science 283: 393-397.

Hwa V, Little B, Kofoed EM, Rosenfeld RG (2004) Transcriptional regulation of insulin-like growth factor-I (IGF-I) by interferon-gamma (IFN-g) requires Stat-5b. J Biol Chem 279: 2728-2736.

Kofoed EM, Hwa V, Little B, Woods K A, Buckway CK., Tsubaki, J, Pratt KL, Bezrodnik L, Jasper H, Tepper A, Heinrich JJ, Rosenfeld RG (2003) Growth-hormone insensitivity (GHI) associated with a STAT-5b mutation. New Engl J Med 349: 1139-1147.

Krempler A, Qi Y, Triplett, AA, Zhu J, Rui H, Wagner KU (2004) Generation of a conditional knockout allele for the Janus kinase 2 (Jak2) gene in mice. Genesis 40: 52-57.

Laron Z, Pertzelan A, Mannheimer S (1966) Genetic pituitary dwarfism with high serum concentation of growth hormone--a new inborn error of metabolism? Isr J Med Sci 2: 152-155.

Leonard WJ, O'Shea JJ (1998) JAKS and STATS: biological implications. Annu Rev Immunol 16: 293-322.

Levy DE, Darnell JE Jr (2002) STATs: transcriptional control and biological impact. Nature Rev Mol Cell Biol 3: 651-662.

Liu JP, Baker J, Perkins AS, Robertson EJ, Efstratiadis A (1993) Mice carrying null mutations of the genes encoding insulin-like growth factor I (Igf-1) and type 1 IGF receptor (Igf1r). Cell 75: 59-72.

Nagy ZS, Wang Y, Erwin-Cohen RA, Aradi J, Monia B., Wang LH, Stepkowski SM, Rui H, Kirken RA (2002).Interleukin-2 family cytokines stimulate phosphorylation of the Pro-Ser-Pro motif of STAT5 transcription factors in human T-cells: resistance to suppression of multiple serine kinase pathways. J Leukoc Biol 72: 819-828.

Neubauer H, Cumano A, Muller M, Wu H, Huffstadt U, Pfeffer K (1998) Jak2 deficiency defines an essential developmental checkpoint in definitive hematopoiesis. Cell 93: 397-409.

Pages G, Guerin, , Grall, D., Bonino F, Smith A, Anjuere F, Auberger P, Pouyssegur J (1999) Defective thymocyte maturation in p44 MAP kinase (Erk1) knockout mice. Science 286: 1374-1377.

Parganas E, Wang D, Stravopodis D, Topham DJ., Marine J, Teglund S, Vanin EF, Bodner S, Colamonici OR, van Deursen JM, Grosveld G, Ihle JN (1998) Jak2 is essential for signaling through a variety of cytokine receptors. Cell 93: 385-395.

Park S-H, Yamashita H, Rui H, Waxman DJ (2001) Serine phosphorylation of GH-activated signal transducer and activator of transcription 5a (STAT5a) and STAT5b: impact on STAT5 transcriptional activity. Mol Endocrinol 15: 2157-2171.

Peng XD, Xu PZ, Chen ML, Hahn-Windgassen A, Skeen J, Jacobs J, Sundararajan D, Chen WS, Crawford SE, Coleman KG, Hay N (2003) Dwarfism, impaired skin development, skeletal muscle atrophy, delayed bone development, and impeded adipogenesis in mice lacking Akt1 and Akt2. Genes Dev 17: 1352-1365.

Pircher TJ, Flores-Morales A, Mui ALF, Saltiel AR, Norstedt G, Gustafson TA, Haldosen L-A (1997) Mitogen-activated protein kinase kinase inhibition decreases growth hormone stimulated transcription mediated by STAT5. Mol Cell Endocrinol 133: 169-176.

Ram PA, Waxman DJ (1999) SOCS/CIS protein inhibition of growth hormone-stimulated STAT5 signaling by multiple mechanisms. J Biol Chem 274: 35553-35561.

Rosenfeld RG, Rosenbloom AL, Guevara-Aguirre J (1994) Growth hormone (GH) insensitivity due to primary GH receptor deficiency. Endocrinol Rev 15: 369-390.

Saba-El-Leil MK, Vella FDJ, Vernay B, Voisin L, Chen L, Labrecque N, Ang S-L, Meloche S (2003) An essential function of the mitogen-activatd protein kinase Erk2 in mouse trophoblast development. EMBO Rep 4: 964-968.

Savage MO, Burren CP, Blair JC, Woods KA, Metherell L, Clark AJ, Camacho-Hubner C (2001) Growth hormone insensitivity: pathophysiology, diagnosis, clinical variation and future perspectives. Horm Res 55: 32-35.

Savage MO, Rosenfeld RG (1999) Growth hormone insensitivity: a proposed revised classification. Acta Paediatr Suppl 428: 147.

Shoba LNN, Newman M, Liu W, Lowe WL (2001) LY294002, an inhibitor of phosphatidylinositol 3-kinase inhibits GH-mediated expression of the IGF-I gene in rat hepatocytes. Endocrinology 142: 3980-3986.

Suzuki H, Terauchi Y, Fujiwara M, Aizawa S, Yazaki Y, Kadowaki T,Koyasu S (1999) Xid-like immunodeficiency in mice with disruption of the p85alpha subunit of phosphoinositide 3-kinase. Science 283: 390-392.

Teglund S, McKay C, Schuetz E, van Deursen JM, Stravopodis D, Wang D, Brown M, Bodner S, Grosveld G, Ihle JN (1998) Stat5a and Stat5b proteins have essential and nonessential, or redundant, roles in cytokine responses. Cell 93: 841-850.

Udy GB, Towers RP, Snell RG, Wilkins RJ, Park SH, Ram PA, Waxman DJ, Davey HW (1997). Requirement of STAT5b for sexual dimorphism of body growth rates and liver gene expression. Proc Natl Acad Sci USA 94: 7239-7244.

Woelfle J, Chia DJ, Rotwein P (2003) Mechanisms of growth hormone (GH) action. Identification of conserved Stat5 binding sites that mediate GH-induced insulin-like growth factor-I gene activation. J Biol Chem 278: 51261-51266.

Woods KA, Camacho-Hubner C, Savage MO, Clark AJ (1996) Intrauterine growth retardation and postnatal growth failure associated with deletion of the insulin-like growth factor I gene. New Engl J Med 335; 1363-1367.

Yamashita H, Xu J, Erwin RA, Farrar WL, Kirken RA, Rui H (1998) Differential control of the phosphorylation state of proline-juxtaposed serine residues Ser[725] of Stat5a and Ser[730] of Stat5b in prolactin-sensitive cells. J Biol Chem 273: 30218-30224.

Yi W, Kim SO, Jiang J., Park SH, Kraft A.S, Waxman DJ, Frank SJ (1996) Growth hormone receptor cytoplasmic domain differentially promotes tyrosine phosphorylation of signal transducers and activators of transcription 5b and 3 by activated JAK2 kinase. Mol Endocrinol 10: 1425-1443.

Growth Hormone Receptor Signaling and Differential Actions in Target Tissues Compared to IGF-I

Paul A. Kelly[1], Anne Bachelot[1],
Athanassia Sotiropoulos[1] and Nadine Binart[1]

Summary

Growth hormone (GH) and IGF-I bind to specific membrane-bound receptors located in widely distributed target tissues. Although initial post-receptor signal transduction pathways differ - GH: associated tyrosine kinase (Jak) that activates signal transducers and activators of transcription (Stat) transcription factors; IGF-I: intrinsic tyrosine kinase that activates insulin receptor substrate (IRS) docking proteins, involved in several down stream effector pathways - many of the pathways are overlapping for GH and IGF-I, which makes it sometimes difficult to determine which hormone is responsible for the action being evaluated.

GH and IGF-I are best known for their stimulatory effects on the growth of bone and soft tissues. Most, but not all, of the known GH actions are mediated by circulating or endocrine IGF-I, produced essentially by the liver. However, many other tissues also synthesize GH and IGF-I locally, and thus each could also function as an autocrine/paracrine growth regulator. Recent in vivo studies using transgenic and classical or tissue-specific knockout models have helped shed light on how the two hormone/growth factors function. GH and IGF-I have independent as well as overlapping functions, and both are needed for maximal effect. However, increasing the circulating levels of IGF-I is frequently sufficient to induce a maximal response.

We examined bone development and remodeling in GH receptor (GHR) knockout (KO) and Stat5ab KO mice (kindly provided by J Kopchick and J Ihle). Markers of bone formation and resorption were reduced in GHR KO mice after two weeks of age. IGF-I treatment almost completely rescued all defects of bone growth and remodeling observed in GHR KO mice. Although bone length is slightly reduced in Stat5ab KO mice, the lack of any effect on trabecular bone remodeling or growth-plate width strongly suggests that the effects of GH in bone may not involve Stat5 activation.

The role of GH and IGF-I on reproductive functions was studied in female GHR KO mice. Litter size was markedly decreased in these animals due to a reduction in the rate of ovulation. IGF-I treatment was ineffective in rescuing this defect, suggesting that the effects of GH on follicular growth are independent of

---

[1] Inserim Unit 584 – Hormone Targets, Faculty of Medicine René Descartes – Paris 5, Paris, France

circulating IGF-I. In the same model, the actions of GH and IGF-I on muscle cell growth and differentiation were studied in vivo and in vitro. The absence of GH signaling resulted in a significant reduction in muscle mass without affecting the fiber number.

Almost all tissues except the liver express IGF-I transcripts in the absence of a functional GHR. Hepatic IGF-I production is dependent on GH, and this *endocrine* IGF-I, working together with GH, is primarily responsible for most of the growth signaling pathways, although this model may not be valid for all GH/IGF-I responsive tissues.

Introduction

Growth hormone (GH), secreted by somatotropic cells of the anterior pituitary, is the hormone primarily responsible for growth of long bones and soft body tissues. Thus, an absence of GH leads to dwarfism, whereas an excess of GH causes acromegaly/gigantism, depending on the age of onset of the adenoma responsible for the hypersecretion of GH. The actions of GH are classified as *direct* and *indirect*: direct actions are those that immediately follow the activation of the GH receptor located at the plasma membrane of the target cell, and include actions on growth of bones and soft tissues, lipolysis, and glucose metabolism. GH also acts directly on the liver to produce insulin-like growth factor -I (IGF-I), a member of the family of growth-promoting polypeptides (Le Roith et al. 2001a). The liver produces large amounts of IGF-I that attain the peripheral circulation and thus IGF-I is considered both a hormone and a local growth factor (Fig. 1).

Receptors and Signaling Pathways

The GH receptor, the cDNA of which was cloned in 1987 (Leung et al. 1987), was the first identified member of the Class 1 Cytokine Receptor Superfamily, which now comprises more than 30 different members. GH binds to a single-pass, membrane bound receptor in its extracellular domain. The three-dimensional (3D) crystal structure of the extracellular domain of the receptor produced the surprising observation that a single molecule of GH binds two molecules of receptor (De Vos et al. 1992). Recent studies suggest that the GH receptor exists as a preformed dimer at the cell surface, and the binding of GH through site 1 of GH to the first receptor and then via site 2 to the second receptor results in a functionally active dimer (Frank 2002). This interaction appears to be the initial event in GH signaling. At present, little or nothing is known about the 3D structure of the cytoplasmic domain of the GH receptor, although a number of signaling pathways have been identified.

Although no kinase domain was found in the cytoplasmic domain of the GH receptor, as had been shown for growth factor receptors, GH was known to induce phosphorylation on tyrosine residues of the receptor as well as of other proteins. In fact, in 1993, an associated tyrosine kinase known as Jak2 was identified as the kinase responsible for GH receptor activation (Argetsinger et al. 1993).

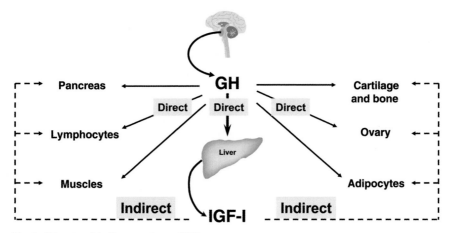

Fig. 1. Direct and indirect actions of GH.

Very rapidly after that, prolactin, erythropoietin, and IL-3 were also shown to activate Jak2. There are a total of four members of this kinase family, Jak1, Jak2, Jak3, and Tyk2.

The role of dimerization of the receptors is to bring into close proximity the two Jak2 molecules, each of which transphosphorylate a tyrosine residue on the other Jak2 molecule, increasing the kinase activity of Jak2. This process leads in turn to the phosphorylation of tyrosine residues in the cytoplasmic domain of the GH receptor (GHR), which acts as a high affinity binding site for a large number of signaling molecules, primarily via Src homology 2 (SH2) domains.

JAK STAT Pathway

GH signaling via the GHR involves the now classical JAK-STAT (signal transducer and activator of transcription) pathway (Herrington and Carter-Su 2001). STAT molecules play a key role in this signaling pathway. STAT molecules bind to phosphorylated tyrosine residues of the cytoplasmic domain and at least four of them are activated: Stats 1, 3, 5A, and 5B. Once activated, they dissociate (via an unknown mechanism) from the receptor, dimerize, and bind to specific recognition sites on the promoters of various target genes.

MAP Kinase Pathway

Next, GH stimulates the MAP (mitogen-activated protein) Kinase Pathway. Jak2 phosphorylates Shc, an adaptor protein, allowing the activation of the Shc-Grb2-Sos-Ras-Mek-MAP (ERK) kinase and regulation of the transcription of genes involved in the cell cycle and in other growth- and differentiation-related events (Herrington and Carter-Su 2001).

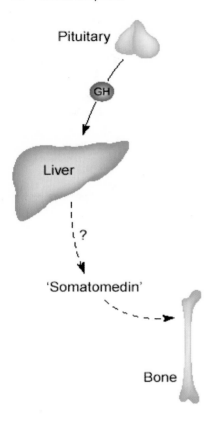

Fig. 2. The original somatomedin hypothesis (after Le Roith et al. 2001b)

Other Pathways

Insulin receptor substrates (IRS) –1 and –2 are docking proteins, first identified as being downstream integrators of signals from the insulin and IGF-I receptors, and are also phosphorylated by GH activation of its receptor. It is believed that many of the metabolic effects attributed to GH are mediated by Jak2 activation and phosphorylation of the IRS proteins (Herrington and Carter-Su 2001).

GH also leads to an increased concentration of intracellular calcium via an influx of extracellular calcium through voltage-dependent L-type calcium channels. The increased calcium levels may be responsible for the period of GH refractoriness. GH also increases intracellular diacylglycerol (DAG), which in turn activates protein kinase-C (PKC) and leads to the generation of inositol triphosphate (IP3). Thus, DAG can be elevated either indirectly or directly by phosphatidylcholine breakdown (Herrington and Carter-Su 2001).

Bone Development and the Somatomedin Hypothesis Revisited

GH is a major regulator of postnatal longitudinal bone growth. IGF-I is expressed in the liver and also in peripheral tissues. The liver is thought to be the major source of circulating IGF-I (Fig. 2); thus it is reasonable to wonder if the circulating or local IGF-I is responsible for growth (Le Roith et al. 2001a). Since there are also direct effects of GH on bone growth, it is important to know whether the circulating IGF-I is necessary or simply additive to the effect of GH. To that end, mouse model systems were developed to better understand the functional importance of local versus circulating IGF-I.

Five years ago, two groups developed a Cre/loxP mouse model in which the IGF-I gene was deleted specifically from the liver, using an interferon or albumin promoter (Sjogren et al. 1999; Yakar et al. 1999). When mice expressing the floxed gene were crossed with interferon or albumin Cre mice, the result was the generation of liver IGF-I-deficient (LID) mice. These mice showed a major (75%) reduction in circulating IGF-I levels, whereas the expression in non-hepatic tissues was normal. Since the LID mice were no different from wild-type littermates with respect to body weight, body length or femoral length, the authors concluded that the normal growth in these animals was mediated by autocrine/paracrine actions of IGF-I combined with non-hepatic sources of circulating IGF-I.

IGFs are bound to high affinity binding proteins (IGFBPs), which are thought to act as carrier proteins, transporting IGFs from the circulation to the target tissues and also slowing down their degradation and thus prolonging their half-life. Among a variety of these binding proteins, IGF-I forms a ternary complex with IGFBP-3 and acid labile subunit (ALS; Le Roith et al. 2001a).

Targeted disruption of the ALS gene resulted in mice showing a 65% reduction in circulating IGF-I and a marked decline in IGFBP levels (Ueki et al. 2000). It was thus surprising that the mice only showed a maximal 10% reduction in body weight. The authors concluded, similar to the conclusions with the LID mice, that locally produced IGF-I would seem to play a crucial role in body growth.

In the elegant study by Yakar et al. (2002), the authors attempted once and for all to clarify the endocrine versus the autocrine/paracrine role of IGF-I in growth and development. They proposed that, by crossing LID with ALS KO mice, the double KOs should have even lower levels of circulating IGF-I. In fact, the double mutant mice were significantly smaller than control, LID or ALSKO animals, having only a slight, but apparently critical, additional reduction in serum IGF-I levels, to 15% of control levels. It thus appears that a threshold level of circulating IGF-I is necessary for normal bone growth. It would seem that the original and revised Somatomedin Hypothesis is again in need of revision (Fig. 3).

A number of studies have attempted to define the precise roles of GH and IGF-I in the regulation of bone growth. A few years ago, we investigated the role of these two hormones on postnatal body growth. We followed the evolution of the growth plate in wild-type and GH GHR KO mice (Sims et al. 2000). We observed that the closing of the growth plate occurred sooner in GHR KO mice than in controls, which accounts for the marked growth retardation observed in these animals (Fig. 4). To see if treatment of the GHR null mice with IGF-I was able to normalize growth, we implanted osmotic minipumps into wild-type and KO

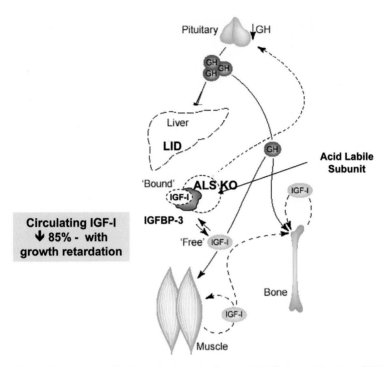

Fig. 3. The somatomedin hypothesis revised again (2003). A combination of IGF-1 knockout (LID) and acid labile subunit (ALS) knockout lowers circulating IGF-1 to a level below the threshold level and thus adouble KO animals have a clear growth defect (after Le Roith et al. 2001b; Yakar et al. 2002).

mice from 2-4, 2-6, and 4-6 weeks of age. Body weight returned to normal in all groups treated with IGF-I, but the animals only attained ~ 85% of body length. Interestingly, chondrocyte proliferation and the width of the growth plate and proliferative zone were fully restored in the GHR KO mice treated until six weeks of age. The fact that IGF-I treatment was not able to counteract the loss of GH signaling in the KO mice could be due to the short treatment periods (four weeks maximum) or to the lack of a direct effect of GH on bone growth that might normally occur in control animals. In support of such a direct effect is that GHR transcripts can clearly be detected in proliferative chondrocytes of the growth plate in wild-type mice. The evolution of the growth plate was also studied in Stat5ab- deficient mice. Although bone length was slightly reduced, we failed to see any reduction in trabecular bone modeling or growth plate width after two weeks of age. Therefore, it would appear that the effects of GH on bone growth after two weeks of age may not be mediated by Stat5.

The landmark work described by Lupu, et al. (2001) seemed to clearly establish the separate and combined roles of IGF-I and GH in postnatal growth in the mouse. This group generated their own like of GHR KO mice and crossbred them with IGF-I KO mice. The resulting double KO (IGF-I + GHR) mouse never

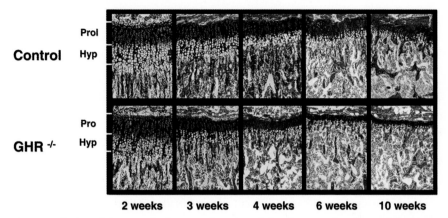

Control

Prol
Hyp

GHR -/-

Pro
Hyp

2 weeks   3 weeks   4 weeks   6 weeks   10 weeks

Fig. 4. Evolution of the growth plate as a function of time in control and growth hormone receptor (GHR) null mice. Closing of the growth plate occurs sooner in GHR knockout (KO) mice, which is responsible for the reduced size of the mutant mice.

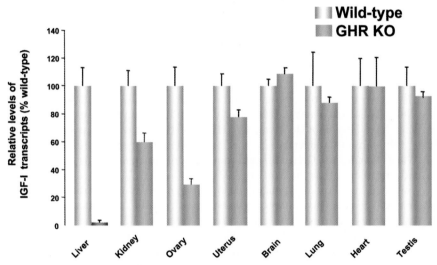

Fig. 5. Relative expression of IGF-1 transcripts measured by semi-quantitative PCR analysis in tissues of GH receptor (GHR) KO mice (after (Lupu et al. 2001)

surpassed a body weight of 5 g and thus is the second smallest known mammal. Evaluation of the various growth curves revealed that 35% of body growth was due to IGF-I alone, 14% to GH alone, and 34% to a combined effect of GH and IGF-I. Growth unrelated to either GH or IGF-I was 17%, that is, the size of the double KO was 17% of that of the wild-type control. Finally, this group measured the relative expression of IGF-I transcripts (by semi-quantitative PCR) in tissues of GHR KO mice. Figure 5 shows that liver IGF-I transcripts are the only ones that are uniquely dependent on GH.

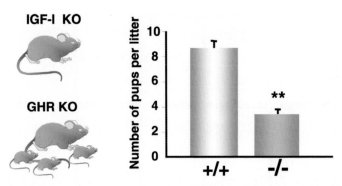

Fig. 6 . The GH–IGF-I axis and fertility. IGF-I KO females are sterile. The ovary and follicles are blocked at a pre-antral stage. GHR null mice are fertile but litter size is reduced (after Bachelot et al. 2002).

Reproductive Functions

GH and IGF-I are involved in reproductive function, especially during sexual maturation. Studies involving IGF-I KO mice have clearly shown that IGF-I is essential for the acquisition of responsiveness to follicle stimulating hormone (FSH), enhancement of FSH actions on granulosa cells, development of follicles beyond the antral stage, and completion of oocyte growth and maturation. Much less is known of the role of GH in ovarian function (Baker et al. 1996; Zhou et al. 1997). In Laron dwarfs (GH insensitivity syndrome; Laron et al. 1966), pregnancies have been reported, so the fertility of the patients is apparently normal, although folliculogenesis has not been extensively examined (Menashe et al. 1991). We used the GHR-deficient mouse model to examine the reproductive phenotype in detail (Bachelot et al. 2002).

As shown in Figure 6, the major reproductive defect is a dramatic decrease in litter size. This decrease is accompanied by a three-fold reduction in the ovulatory response to exogenous gonadotropin treatment. Thus, the reduced rate of ovulation is due to an ovarian defect rather than a deficiency of pituitary godadotropins. Histological examination of sections of ovaries from GHR KO mice clearly shows that the number of follicles per ovary is markedly reduced, but all categories of follicles are represented. The process of implantation occurs normally. Interestingly, the number of healthy follicles from antral and preovulatory stages is markedly decreased in GHR KO compared to wild-type mice. The binding of radiolabeled LH , FSH and IGF-I to ovaries from wild-type and KO mice was similar, suggesting the fully functional receptors remain present on the functioning ovary. Finally, since the defect could be due to a lack of IGF-I, animals were treated with growth factor in osmotic mini-pumps for two weeks. Although the treatment was effective, it failed to rescue either fertility or ovarian responsiveness to gonadotropins. It therefore appears that the effect of GH in the ovary is independent of IGF-I. GH deficiency in female mice is thus responsible for reduced litter size due to a reduced ovulation rate,

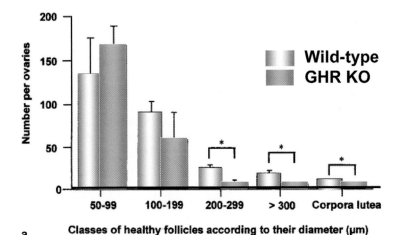

a   **Classes of healthy follicles according to their diameter (μm)**

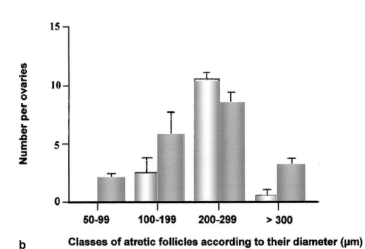

b   **Classes of atretic follicles according to their diameter (μm)**

Fig.7. Classes of healthy (A) and atretic (B) follicles based on their diameter. Follicles were ranked by size as indicated and the degree of atresia by morphometric analysis of ovaries of wild-type and GHR KO mice. Values are presented as means ± SEM. *, $p < 0.05$ (after Bachelot et al. 2002).

reduced number of antral follicles and increased rate of atretic/healthy follicles >200 μm (Fig. 7).

Muscle Growth And Differentiation

Skeletal muscle can increase in size by three approaches: 1) an increase in the number of muscle fibers; 2) an increase in size due to fiber fusion, with more nuclei per cell than for the precursor; and 3) hypertrophy, with a larger mass, but with the same number of nuclei per fiber being maintained in the cells.

Muscle size is reduced in GHR-deficient mice (Sotiropoulos et al., manuscript in preparation). Since the overall body size of GHR KO mice was smaller, muscle mass was also calculated as a proportion of body weight. For three different muscles, soleus, EDL and tibialis anterior, the reductions were all highly significant, suggesting that the reduction in mass was significant.

Growth hormone is able to affect myotube size in vitro. Myotubes from wild-type and GHR KO mice were incubated in the absence or presence of GH. Both cell size and the number of nuclei were reduced, confirming data obtained in vivo. It thus appears that GH regulates myotube size by facilitating muscle cell fusion. GHR-deficient mice show a reduction in muscle fiber size without any change in number, again confirming that the effect is due to a fusion defect.

Since muscle cells contain both GH and IGF-I receptors, it was important to determine if the effect of GH on myotube growth was a direct effect or mediated by IGF-I. IGF-I mRNA expression was evaluated in vitro in the absence and presence of GH. No correlation was observed between GH-induced hypertrophy and IGF-I mRNA expression measured by semi-quantitative RT-PCR. In addition, conditioned medium from myotubes grown in the presence of GH were unable to stimulate the growth of myotubes from GHR KO mice. Thus, GH-induced hypertrophy is likely to be IGF-I independent and thus a direct effect of GH.

References

Argetsinger LS, Campbell GS, Yang X, Witthuhn BA, Silvennoinen O, Ihle JN, Carter-Su C (1993) Identification of JAK2 as a growth hormone receptor-associated tyrosine kinase. Cell 74: 237-244

Bachelot A, Monget P, Imbert-Bollore P, Coschigano K, Kopchick JJ, Kelly PA, Binart N (2002) Growth hormone is required for ovarian follicular growth. Endocrinology 143: 4104-4112

Baker J, Hardy MP, Zhou J, Bondy C, Lupu F, Bellve AR, Efstratiadis A (1996) Effects of an Igf1 gene null mutation on mouse reproduction. Mol Endocrinol 10: 903-918

De Vos AM, Ultsch M, Kossiakoff AA (1992) Human growth hormone and extracellular domain of its receptor: crystal structure of the complex. Science 255: 306-312

Frank SJ (2002) Receptor dimerization in GH and erythropoietin action--it takes two to tango, but how? Endocrinology 143: 2-10

Herrington J, Carter-Su C (2001) Signaling pathways activated by the growth hormone receptor. Trends Endocrinol Metab 12: 252-257

Laron Z, Pertzelan A, Mannheimer S (1966) Genetic pituitary dwarfism with high serum concentation of growth hormone--a new inborn error of metabolism? Isr J Med Sci 2: 152-155

Le Roith D, Bondy C, Yakar S, Liu JL, Butler A (2001a) The somatomedin hypothesis: 2001. Endocr Rev 22: 53-74

Le Roith D, Scavo L, Butler A (2001b) What is the role of circulating IGF-I? Trends Endocrinol Metab 12: 48-52

Leung DW, Spencer SA, Cachianes G, Hammonds RG, Collins C, Henzel WJ, Barnard R, Waters MJ, Wood WI (1987) Growth hormone receptor and serum binding protein: purification, cloning and expression. Nature 330: 537-543

Lupu F, Terwilliger JD, Lee K, Segre GV, Efstratiadis A (2001) Roles of growth hormone and insulin-like growth factor 1 in mouse postnatal growth. Dev Biol 229: 141-162

Menashe Y, Sack J, Mashiach S (1991) Spontaneous pregnancies in two women with Laron-type dwarfism: are growth hormone and circulating insulin-like growth factor mandatory for induction of ovulation? Human Reprod 6: 670-671

Sims NA, Clement-Lacroix P, Da Ponte F, Bouali Y, Binart N, Moriggl R, Goffin V, Coschigano K, Gaillard-Kelly M, Kopchick J, Baron R, Kelly PA (2000) Bone homeostasis in growth hormone receptor-null mice is restored by IGF-I but independent of Stat 5. J Clin Invest 106: 1095-1103

Sjogren K, Liu JL, Blad K, Skrtic S, Vidal O, Wallenius V, LeRoith D, Tornell J, Isaksson OG, Jansson JO, Ohlsson C (1999) Liver-derived insulin-like growth factor I (IGF-I) is the principal source of IGF-I in blood but is not required for postnatal body growth in mice. Proc Natl Acad Sci U S A 96: 7088-7092

Ueki I, Ooi GT, Tremblay ML, Hurst KR, Bach LA, Boisclair YR (2000) Inactivation of the acid labile subunit gene in mice results in mild retardation of postnatal growth despite profound disruptions in the circulating insulin-like growth factor system. Proc Natl Acad Sci USA 97: 6868-6873

Yakar S, Liu JL, Stannard B, Butler A, Accili D, Sauer B, LeRoith D (1999) Normal growth and development in the absence of hepatic insulin-like growth factor I. Proc Natl Acad Sci USA 96: 7324-7329

Yakar S, Rosen CJ, Beamer WG, Ackert-Bicknell CL, Wu Y, Liu JL, Ooi GT, Setser J, Frystyk J, Boisclair YR, LeRoith D (2002) Circulating levels of IGF-1 directly regulate bone growth and density. J Clin Invest 110: 771-781

Zhou J, Kumar TR, Matzuk MM, Bondy C (1997) Insulin-like growth factor I regulates gonadotropin responsiveness in the murine ovary. Mol Endocrinol 11: 1924-1933

IGF-I and Brain Growth: Multifarious Effects on Developing Neural Cells and Mechanisms of Action

Teresa L. Wood[1], Terra J. Frederick[1] and Jennifer K. Ness[1]

Summary

Numerous investigators have provided data supporting an essential role for IGF-I in growth of the brain. IGF-I contributes to multiple processes during brain development, including neural cell survival, proliferation, differentiation and maturation. The IGF type I receptor (IGF-IR) is present on all cell types in the brain, and IGF-I has known actions on neural stem and progenitor cells as well as neurons and glia. IGF-I is highly expressed throughout the brain during development, and its expression is retained in the meninges and in many cell types in the adult brain. While IGF-I has multiple actions on developing neural cells, very few studies have addressed the mechanisms or pathways by which IGF-I mediates these multiple effects. The goal of this chapter is to briefly review data on IGF-I in the developing brain and then to discuss more recent studies that focus on the mechanisms for its varied actions.

Introduction

Evidence from transgenic and gene-targeted mouse lines has provided clear support for a role for IGF-I in growth and development of the brain. Moreover, emerging data from human mutations in IGF-I and the IGF-IR further support an essential role for IGF signaling in normal brain maturation and function. Both IGF-I and its primary signaling receptor, the IGF-IR, are expressed throughout the developing central nervous system (CNS; Baron-Van Evercooren et al. 1991; Bartlett et al. 1992; Bondy et al. 1990, 1992; Wilkins et al. 2001). IGF-I has known actions on neural stem and progenitor cells as well as on developing and adult neurons and glia. IGF-I has pleiotropic effects on neural cells; there is evidence to support a role for IGF-I in proliferation and cell fate decisions in neural stem and progenitor cells, and in survival, differentiation and maturation of both immature and mature neurons and glia. Until recently, little was known about the mechanisms or pathways by which IGF-I promotes these diverse effects on cells of the CNS. In this review, we will summarize data on the role of IGF-I in brain

---

[1] Department of Neural & Behavioral Services, Penn State College of Medicine, 500 University Drive, Hershey, PA 17033

Carel et al.
Deciphering Growth
© Springer-Verlag Berlin Heidelberg

growth and maturation and discuss recent reports on the mechanisms and pathways for its diverse actions on developing neural cells.

IGF-I Transgenic and Gene-Targeted Mice

The initial in vivo evidence for a specific role for IGF-I in brain growth came from analyses of transgenic mice (for reviews, see D'Ercole et al. 1996, 2002; Wood 1995). In initial experiments, Mathews and colleagues (1988) established transgenic mouse lines that overexpressed IGF-I from the metallothionine promoter (MT-IGF-I), resulting in increased body size as well as growth of the brain. When the MT-IGF-I transgenic mice were crossed with a growth hormone (GH)-deficient mouse line generated by expression of a a diptheria toxin gene from the GH promoter, it was further demonstrated that the increased brain growth in the IGF-I transgenic mice was maintained in the absence of GH-producing cells (Behringer et al. 1990). Moreover, MT-GH transgenic mice have increased body growth and selective organ overgrowth, but no increase in brain size (Palmiter et al. 1982, 1983). Taken together, these data support a role for IGF-I in brain growth, independent of its actions as a mediator of GH in somatic growth.

Subsequent to the studies establishing that IGF-I can promote brain growth, more detailed analyses of both IGF-I overexpressing and IGF-I gene-targeted mice provided evidence that IGF-I has diverse actions on neural cells in the developing brain (for reviews, see D'Ercole et al. 1996, 2002; Wood 1995). Overexpression of IGF-I results in a 55% increase in brain growth, due to an increase in cell size, increased numbers of both neurons and oligodendrocytes, and increased myelin content (Carson et al. 1993; Ye et al. 1995).

Definitive evidence that IGF-I is essential for brain growth was demonstrated from analysis of brains from mice carrying a null mutation in the *Igf1* gene (Beck et al. 1995). Absence of IGF-I reduces the numbers of neurons, axons, oligodendrocytes and myelin content. Interestingly, the reduction in numbers of myelinated axons is proportionally greater than the reduction in numbers of unmyelinated axons. The authors concluded that IGF-I is necessary for axon growth and maturation during CNS myelination, in addition to regulating neuron and oligodendrocyte numbers. In addition, loss of IGF-I affects specific neuronal populations differentially, such that some neurons survive whereas other classes of neurons die or fail to develop in the absence of IGF-I (Beck et al. 1995; Cheng et al. 1998). One study suggested that loss of oligodendrocytes in the IGF-I null brains is secondary to loss of neurons in specific brain regions (Cheng et al. 1998). However, a recent analysis of the IGF-I null mice at earlier times in development supports the hypothesis that IGF-I also directly regulates oligodendrocyte development and myelination, but that other mechanisms, such as increased levels of IGF-II, compensate for the loss of IGF-I by adult ages (Ye et al. 2002).

Diverse functions of IGF-I on multiple neural cell types

Stem cells and neural progenitors

IGF-I is well known as a survival factor for neural cells of all stages. In addition to its activation by IGF-I, the IGF-I receptor (IGF-IR) is activated by insulin in the micromolar range (LeRoith et al. 1995), concentrations commonly used in chemically defined culture media for many primary cells, including neural progenitors as well as neurons and oligodendroglia (Bottenstein et al. 1980; McCarthy and de Vellis 1980). In most cases, stimulation of the IGF-IR is required for survival of neural cells in vitro. Thus, it has been difficult to accurately distinguish other actions of IGF-I on neural cells except in short-term assays.

In vitro studies from several laboratories provided initial support for the hypothesis that IGF-I promotes proliferation of neural stem and progenitor cells. IGF-I promotes DNA synthesis in neuronal precursors from embryonic day 15-16 (e15-16) mouse brains (Lenoir and Honegger 1983) and in cultured rat sympathetic neuroblasts of the developing peripheral nervous system (DiCicco-Bloom and Black 1988). In another study, Drago et al. (1991) demonstrated that neuroepithelial progenitor cells from e10 mouse brain produce IGF-I and are dependent on it for their survival. In contrast to the previous reports, IGF-I by itself had no mitogenic activity on the p10 neural progenitors; however, it enhanced mitogenic activity of fibroblast growth factor-2 (FGF-2) in short-term assays (Drago et al. 1991). A recent study similarly supports a role for IGF-I in augmenting FGF-2 mediated proliferation of adult hippocampal stem/progenitor cells (Aberg et al. 2003); however, analysis of embryonic striatal neural stem cells suggests that IGF-I enhances EGF-mediated proliferation of these cells (Arsenijevic et al. 2001). The conclusion from the in vitro studies is that IGF-I enhances proliferation of neural stem/progenitor populations either directly or by amplifying the actions of other mitogens. Recently, support for IGF-I in proliferation of neural stem/progenitor cells in vivo was provided from a transgenic mouse model, where IGF-I is expressed from nestin regulatory elements active in neural stem and progenitor populations. The increased IGF-I in the embryonic neuroepithelium results in an increase in BrdU-labeled cells at e14, when there are very low levels of endogenous cell death. Moreover, the increased neuroepithelial proliferation in embryonic stages in the nestin-IGF-I transgenic brains correlates with increased brain size and numbers of neurons by late gestation and early postnatal ages (Popken et al. 2004). Importantly, neural stem/progenitor cells express the IGF-IR, and there is abundant IGF-I in the developing brain and cerebral spinal fluid during embryogenesis. Taken together, these data support the hypothesis that IGF-I mediates proliferation of early neural stem and progenitor cells during CNS development, independent of its actions as a survival factor.

IGF-I recently has been implicated in mediating cell fate decisions of multipotent neural stem/progenitor cells. Arsenijevic and Weiss (1998) demonstrated that IGF-I promotes differentiation of post-mitotic neuronal precursors, in the absence of any effect on proliferation or survival. Similarly, endogenously produced IGF-I is required for neuronal and glial differentiation from embryonic olfactory stem cells (Vicario-Abejon et al. 2003). Several groups investigating multipotent neural progenitor cells isolated from adult brain have suggested that IGF-I is instructive in neural progenitor cell fate decisions (Aberg et al. 2003; Brooker et al. 2000; Hsieh et al. 2004). Aberg et al. (2003) demonstrated that IGF-I treatment of adult hippocampal progenitors increases the proportion of cells expressing neuronal markers. Endogenously produced IGF-I appears to be important for neuronal differentiation from multipotent cells from the adult forebrain (Brooker et al. 2000). In contrast, Hsieh and colleagues (2004) provided evidence that IGF-I instructs adult hippocampal progenitors to become oligodendrocytes, with a smaller effect on increasing the numbers of neurons.

Neuronal survival

As discussed above, stimulation of the IGF-IR is required for optimal basal survival of most neural cells in culture, including immature neurons. In IGF-I transgenic mice, increased growth in the cerebellum is, in part, the result of decreased apoptosis (Chrysis et al. 2001; Ye et al. 1996). Moreover, in the nestin-IGF-I transgenic mice, which have increased proliferation in the embryonic neuroepithelium, expression of IGF-I also reduces cell death in cortical regions during postnatal ages (Popken et al. 2004). Thus, the increased brain growth observed in these mice is due both to increased proliferation and survival. In vitro, IGF-I protects immature cerebellar granule cells from apoptosis due to withdrawal of serum and depolarizing potassium (Linseman et al. 2002). Similarly, IGF-I protects embryonic dorsal root ganglion neurons from trophic factor withdrawal and hyperosmotic stress (Russell and Feldman 1999; Russell et al. 1998) and embryonic motor neurons from glutamate toxicity (Vincent et al. 2004). IGF-I also protects mature neurons from apoptotic death due to axotomy (Kermer et al. 2000) or hypoxia-ischemia in vivo (Gluckman et al. 1992; Guan et al. 1993; Liu et al. 2001a,b).

Oligodendroglia

Cells at all stages of oligodendrocyte development express the IGF-IR (McMorris and McKinnon 1996; McMorris et al. 1993, 1986). In addition, oligodendroglia express IGF-I, particularly in the progenitor stages in vitro (Shinar and McMorris 1995). Consistent with these observations, IGF-I is expressed in the subventricular zone during early postnatal development, a region with high numbers of oligodendrocyte progenitors (OPs; Bartlett et al. 1992). Expression of IGF-I by OP cells enhances cortical neuroblast survival in oligodendrocyte/neuron cocultures (Wilkins et al. 2001), and autocrine/paracrine production of IGF-I has been proposed to regulate oligodendrocyte development (Baron-Van Evercooren et al. 1991). Conversely, IGF-I production by neurons and axons may provide important survival and differentiation signals to oligodendrocytes during myelination.

Numerous in vitro studies have provided considerable data suggesting that IGF-I has multiple roles in oligodendrocyte development, including enhancing proliferation, survival and maturation of oligodendroglia. IGF-I was originally thought of primarily as a survival factor for OPs in vitro (Barres et al. 1992a,b; Barres et al. 1993). While subsequent studies suggested that physiological levels of IGF-I promote DNA synthesis in primary cultures of OPs (McMorris et al. 1993), IGF-I by itself is a weak mitogen for OPs in contrast to platelet-derived growth factor (PDGF) and FGF-2 (Jiang et al. 2001). However, our previous studies demonstrated that maximal mitogenic actions of FGF-2 and PDGF require the presence of IGF-I or of supraphysiological levels of insulin to co-stimulate the IGF-IR (Jiang et al. 2001). An interesting finding from these studies was that the combination of IGF-I and PDGF is additive whereas IGF-I and FGF-2 are synergistic in promoting DNA synthesis in OP cells. These results are consistent with reports that IGF-I and FGF-2 cooperate to promote proliferation of embryonic neural progenitors, suggesting that this is a common role for IGF-I in brain growth.

A role for IGF-I in proliferation of the OP has not been investigated in vivo. In the MT-IGF-I overexpressing mouse lines, transgene expression is initiated too late postnatally to significantly alter early OP proliferation, which is maximal in early postnatal ages (Carson et al. 1993; Ye et al. 1995). Similarly, the mouse line carrying a germ-line deletion of IGF-I is not a good model in which to assess a specific role for IGF-I in OP proliferation in vivo. While it is clear that there is a reduction in myelin and oligodendrocyte numbers in the brains of the IGF-I null mice, the phenotype of the IGF-I null mutant mice is complex and includes decreased fetal growth, reduced postnatal viability and loss of specific neuron subtypes in the brain (Beck et al. 1995; Cheng et al. 1998). Thus, it is difficult to clearly establish whether a reduction in oligodendrocyte generation is a primary or secondary defect in the IGF-I null mutants or whether it is due to effects on proliferation, survival or maturation of the cells. Indeed, whether the reduction in oligodendrocytes in the absence of IGF-I is a primary defect or is due secondarily to loss of neurons has been controversial (Cheng et al. 1998; Ye et al. 2002). However, recent analysis of the IGF-I null mutant by Ye et al, (2002) demonstrated a specific reduction in OPs and differentiating oligodendrocytes during postnatal ages during the time when oligodendrocytes are generated. In contrast, and as previously reported, the proportion of oligodendrocytes in the adult IGF-I KO brain is consistent with the number of neurons in wild-type brains (Cheng et al. 1998; Ye et al. 2002). Moreover, Ye et al (2002) suggest that the recovery of oligodendrocytes and myelination in the adult brain is due, in part, to compensation by an elevation of IGF-II expression in the IGF-I knockout brains.

Survival of oligodendroglia

In addition to enhancing basal survival in vitro, IGF-I enhances survival of oligodendroglia from toxic stimuli, including excitotoxicity and cytokine-mediated toxicity. Tumor necrosis factor–alpha (TNF-α), a cytokine implicated in demyelinating disorders, induces apoptosis of OPs and differentiated oligodendrocytes (Ye and D'Ercole 1999). Treatment of these cultures with IGF-I inhibits TNF-α-induced apoptosis (Ye and D'Ercole 1999). Glutamate, an agent implicated in various CNS disorders, induces apoptosis of OPs in a time- and dose-dependent manner. In particular, the late stage of OP, present in high numbers in the prenatal human and perinatal rodent brain, is particularly susceptible to glutamate toxicity (Back et al. 1998; McDonald et al. 1998). Glutamate toxicity is a major mediator of hypoxic-ischemic death of both neurons and glia. In premature human infants, hypoxia-ischemia (H/I) leads to the death of the OP cells that are abundant in the developing white matter at this stage of development. In our recent studies, we determined that IGF-I prevents glutamate-mediated apoptosis in the late OPs (Ness et al. 2004; Ness and Wood 2002). Interestingly, IGF-I levels decrease in the immature brain immediately following H/I (Lee et al. 1996), suggesting the possibility that the decrease in IGF-I concurrent with elevated levels of glutamate contributes to death of the OPs following H/I. Consistent with the ability of IGF-I to protect oligodendroglia from cytokine and glutamate toxicity, IGF-I overexpression in transgenic mice protects mature oligodendrocytes from apoptosis during cuprizone-induced demyelination (Mason et al. 2000).

Oligodendrocyte maturation

In vitro experiments initially suggested that IGF-I enhances the differentiation of early progenitor cells to generate oligodendrocytes (McMorris et al. 1986; McMorris and Dubois-Dalcq 1988). Early OPs isolated from perinatal rat cerebrum and treated with exogenous IGF-I generated increased numbers of oligodendrocytes, based on morphological analysis and expression of oligodendrocyte antigens (McMorris et al. 1986; McMorris and Dubois-Dalcq 1988). Similarly, in human cell cultures, IGF-I stimulation of early human fetal OP cells increases the relative numbers of differentiating oligodendrocytes to late OPs without affecting proliferation (Armstrong et al. 1992; Satoh and Kim 1994).

While it is difficult to rule out the effects of IGF-I on survival in the differentiation experiments, in vitro and in vivo studies also support a role for IGF-I in maturation and myelin production in oligodendrocytes. IGF-I treatment of aggregate cultures isolated from fetal rat brain causes significant increases in the numbers of oligodendrocytes as well as the synthesis and accumulation of myelin (Mozell and McMorris 1991). Similarly, explants of developing mouse spinal cord exposed to IGF-I show increased amounts of myelin and of oligodendrocyte-specific antigens compared to control cultures (Roth et al. 1995). The analysis of the transgenic mice further supports a role for IGF-I in enhancing myelination. As discussed previously, the percentages of myelinated axons and myelin content increase with overexpression of IGF-I whereas the numbers of oligodendrocytes normalized to total cell number are similar in the adult brains of these animals. Thus the IGF-I induced increase in myelin is likely the result of more myelin per oligodendrocyte (Carson et al. 1993; Ye et al. 1995). Additionally, overexpression of IGF-I enhances levels of proteolipid protein and myelin basic protein mRNA by 200%, indicating that IGF-I stimulates expression of myelin protein genes (Ye et al. 1995).

Mechanisms and Signaling Pathways
for IGF-I Actions on Neural Cells

The data reviewed above demonstrate that IGF-I promotes brain growth through actions on neural stem and progenitor cells and on developing neurons and glia. In addition, IGF-I continues to affect neural cell function in the adult CNS and has a role in brain injury. The actions of IGF-I in promoting brain growth include regulation of proliferation, survival, cell fate/differentiation and maturation. Defining the mechanisms and pathways for the multiple functions of IGF-I on different neural cell types is one of the most interesting new areas of research into CNS IGF biology. In the next sections, we discuss recent data on the mechanisms for IGF-I-mediated proliferation and survival of neural cells and on the current understanding of the signaling pathways responsible for these actions.

IGF-I Regulation of Cell Cycle Progression in Neural Cells

In OPs, IGF-I with FGF-2 is responsible for recruiting additional progenitor cells past the G_1-S transition and into S phase. Further analysis of cell cycle kinetics in OP cells has provided information suggesting that cells exposed to the combination of IGF-I/FGF-2 have enhanced progression through the G_1 phase of the cell cycle and past the G_1-S transition point (Fig. 1; Frederick and Wood 2004). Moreover, in addition to enhancing G_1 progression, IGF-I is required for FGF-2 to promote G_2-M progression in the OPs (Fig 1; Frederick and Wood 2004).

Further analyses revealed complex mechanisms for IGF-I and FGF-2 regulation of G_1 progression in OP cells (Frederick and Wood 2004). IGF-I and FGF-2 coordinately enhance expression of cyclin D1 early in G_1 (Fig. 2). Both the levels and the rate of cyclin D1 induction are increased in the presence of the two factors, which reflects the increased rate of G_1 progression. IGF-I has additional effects on the cell cycle machinery during G_1 including stabilization of cyclin E and enhancement of cyclin E/cdk2 complex formation, the activation of which is required for progression past the G_1-S transition (Fig. 3A,B). The combination of IGF-I with FGF-2 also reduces protein levels of the cell cycle inhibitor, p27(Kip1) (Fig. 3A). The net result of these alterations is that the IGF-I/FGF-2-treated cells have significantly enhanced cyclin E/cdk2 activity (Fig. 3C), which is proportional to hyperphosphorylation of the retinoblastoma protein and S-phase entry (Frederick and Wood 2004).

As discussed previously, recent studies demonstrated that overexpression of IGF-I in the embryonic neuroepithelium enhances proliferation of ventricular zone cortical progenitors in transgenic mice (Popken et al. 2004). Further analysis of these mice demonstrated that the increased levels of IGF-I decrease the G_1 phase of the cell cycle in the cortical progenitor cells (Hodge et al. 2004), similar to what we observed in the OPs (Frederick and Wood 2004). Interestingly, FGF-2 is expressed during these ages, and loss of FGF-2 decreases proliferation of ventricular zone progenitors in FGF-2 knock-out mice (Vaccarino et al. 1999). Thus, it is possible that induction of IGF-I in the nestin-IGF-I mice reflects a role for IGF-I in combination with FGF-2 in enhancing cell cycle progression of the embryonic neural progenitor cells, similar to the OP cells. Taken together

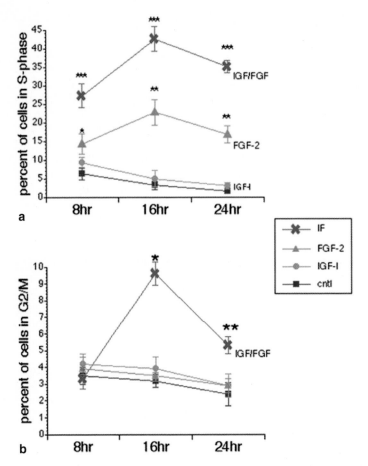

Fig. 1. Results of flow cytometric analysis of OP cells treated with IGF-I and/or FGF-2. OP cell cultures were growth-arrested and treated with IGF-I, FGF-2, IGF-I/FGF-2 (10 ng/ml each), or no growth factors for 8, 16, or 24 hours. Flow cytometric dot plots were generated from cells stained for 7-AAD and BrdU, and quadrant markers were applied. The percentage of cells in S phase (a) and G_2/M (b) after 8, 16, and 24 hours in control and growth factor-treated OP cultures was quantified using the DNA staining analysis program ModFit™. The data represent the mean ± SEM (n = 7 from three separate experiments). (a) * p ≤ 0.02 vs. control, ** p ≤ 0.002 vs. control and IGF-I, *** p ≤ 0.001 vs. control, IGF-I, and FGF-2. (b) * p ≤ 0.001 vs. control, IGF-I, and FGF-2, ** p ≤ 0.01 vs. control, IGF-I, and FGF-2. (Reprinted from Frederick and Wood 2004 with permission from Elsevior).)

with the in vitro studies reviewed previously, which show cooperation of IGF-I and FGF-2 in proliferation of embryonic neural progenitors, it is reasonable to hypothesize that the combination of IGF-I and FGF-2 is important for cell cycle progression of both early neural progenitors that give rise to cortical neurons and of the later OPs that produce oligodendroglia in the postnatal brain. Interestingly, data from the nestin-IGF-I mice suggested no effect of IGF-I on

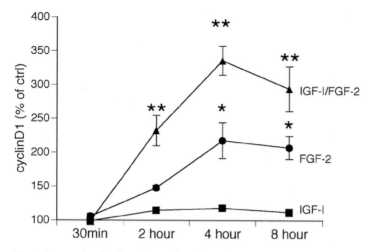

Fig. 2. Expression levels and rate of cyclin D1 expression in OP cells treated with IGF-I and/or FGF-2. Analysis of cyclin D1 from Western immunoblots following growth factor treatment of OP cells between 0 and 8 hours. Cyclin D1 levels were normalized to β-actin levels and are represented as percentage of t_0 control levels. Values represent the mean ± SEM for each condition (n = 3). (A) * p ≤ 0.01 vs. control and IGF-I, ** p ≤ 0.002 vs. control and IGF-I and p ≤ 0.02 vs. FGF-2 alone. (Reprinted from Frederick and Wood 2004 with permission from Elsevior).

G_2-M progression of the neuroepithelial progenitors (Hodge et al. 2004). The difference between this result and our results on OPs might reflect differences in progenitor cell type but more likely is due to the presence of sufficient IGF-I in the in vivo environment to promote G_2-M progression, such that additional IGF-I had no further effect.

IGF-I mediated survival pathways in neural cells

There is accumulating evidence that the mechanisms for IGF-I-mediated protection of neural cells from apoptosis differ between neurons and oligodendroglia. In neurons, IGF-I can inhibit calcium entry and enhance intracellular calcium recovery following exposure to glutamate (Cheng and Mattson 1991, 1992, 1994; Cheng et al. 1993; Mattson and Cheng 1993; Mattson et al. 1993). In contrast, IGF-I does not alter calcium entry or recovery in the late OPs exposed to excess glutamate (Ness et al. 2004). In neurons, IGF-I also induces anti-apoptotic proteins such as Bcl-2 and reduces levels of pro-apoptotic proteins such as Bax and Bim (Baker et al. 1999; Chrysis et al. 2001; Linseman et al. 2002; Matsuzaki et al. 1999; Parrizas and LeRoith 1997; Tamatani et al. 1998). In OP cells exposed to glutamate, IGF-I does not induce protein levels of either Bcl-xL or Bcl-2 or reduce levels of Bax or Bim (Ness et al. 2004). In the late OPs, IGF-I prevents mitochondrial dysfunction, release of cytochrome c and the subsequent cleavage of caspases 9 and 3, predominantly through blocking Bax translocation to the mitochondria (Ness et al. 2004).

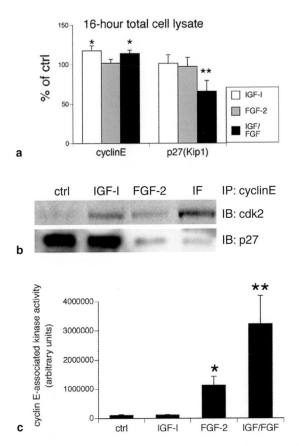

Fig. 3. Regulation of late G_1 complex activation. (a-c) OP cells were treated with growth factors for 8 or 16 hours after growth arrest. (a) Levels of cyclin E and p27(Kip1) are represented as percentage of 16-hr control following normalization to β-actin. (b) Immunoblot analyses for cdk2 and p27(Kip1) following immunoprecipitation with antibodies to cyclin E from 16-hour growth factor-treated and control lysates to analyze late G_1 complex association. (c) Graph of cyclin E-associated kinase activity determined using histone H1 as a substrate following cyclin E immunoprecipitation of 16-hour growth factor-treated and untreated OP cell lysates. (a) * $p \leq 0.05$ vs. control and FGF-2, ** $p \leq 0.05$ vs. control, IGF-I, and FGF-2, (c) * $p \leq 0.03$ vs. control and IGF-I, ** $p \leq 0.001$ vs. control and IGF-I, and $p \leq 0.02$ vs. FGF-2. (Reprinted from Frederick and Wood 2004 with permission from Elsevier).)

The PI3-Kinase Pathway and IGF Actions in the Brain

The PI3-Kinase (PI3K) pathway is a major signal transduction pathway for survival in many cell types, including neurons and glia. Not surprisingly, IGF-I, like many other trophic factors, utilizes this pathway to promote survival of neural cells. Activation of PI3K and its downstream mediator, Akt, is obligatory for survival of oligodendroglia (Flores et al. 2000; Ness and Wood 2002; Vemuri and

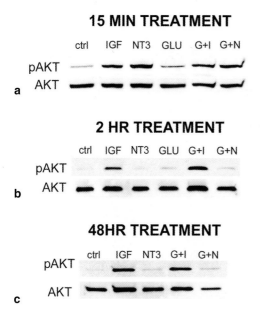

15 MIN TREATMENT

ctrl   IGF   NT3   GLU   G+I   G+N

a   pAKT

AKT

2 HR TREATMENT

ctrl   IGF   NT3   GLU   G+I   G+N

b   pAKT

AKT

48HR TREATMENT

ctrl   IGF   NT3   G+I   G+N

c   pAKT

AKT

Fig. 4. IGF-I sustains phosphorylation of Akt through 48 hours while NT3 transiently activates Akt. Late OPs were treated with trophic factors and Akt phosphorylation was assessed by Western immunoblotting on the isolated proteins. Blots were stripped and used for analysis of total Akt levels. (a-c) Representative immunoblots of Akt phosphorylation (pAKT) and total Akt (AKT) in late OP cultures treated with glutamate in the presence or absence of IGF-I and NT-3 after 15 min (a), 2 hours (b), or 48 hours (c). (Reprinted from Ness and Wood 2002 with permission from Elsevier).)

McMorris 1996). IGF-I also utilizes the PI3K/Akt pathway to promote survival and block apoptotic death in neurons (Kermer et al. 2000; Leinninger et al. 2004; Linseman et al. 2002; Vincent and Feldman 2002; Vincent et al. 2004; Zheng and Quirion 2004).

In OP cells, IGF-I is a potent activator of PI3K and Akt (Ness and Wood 2002). Unlike neurotrophin-3 (NT-3), also a survival factor for OPs, IGF-I sustains phosphorylation of Akt (Fig. 4), which correlates with its ability to provide long-term protection of these cells from apoptosis due to either trophic factor deprivation or excitotoxicity (Ness et al. 2002, 2004; Ness and Wood 2002). The transient activation of Akt by NT-3 in OP cells is correlated with a rapid down-regulation of activation and total levels of TrkC, its primary signaling receptor (Ness and Wood 2002). These results suggest that NT-3 binding to the TrkC receptor results in its down-regulation and ultimately in loss of survival signaling. Similar reports in neurons on other neurotrophin receptors support the hypothesis that this is a general mechanism by which neurotrophins regulate their own actions (Frank et al. 1996; Knusel et al. 1997; Zhang et al. 2000). In contrast, levels of IGF-IR phosphorylation and total receptor levels are stable in OP cells exposed to IGF-I (Fig. 5).

The link between PI3K/Akt activation and interference with the mitochondrial death pathway is still under investigation in many cell types. IGF-I promotes survival against trophic factor deprivation by stimulating Akt and blocking activation of the forkhead transcription factor, FKHRL1, in cerebellar granule neurons and embryonic hippocampal neurons (Linseman et al. 2002; Zheng and Quirion 2004). In the cerebellar granule neurons, this pathway results in suppression of the pro-apoptotic protein, Bim (Linseman et al. 2002). In OP cells,

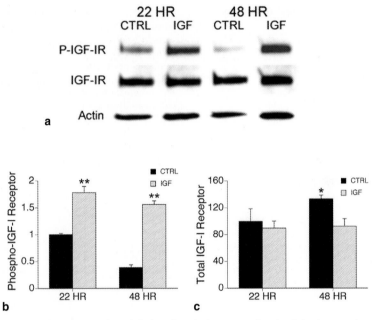

Fig. 5. IGF-IR expression and phosphorylation are maintained during continuous exposure to IGF-I through 48 hours. Late OPs were treated with or without IGF-I, and isolated protein was analyzed for IGF-IR phosphorylation and for total IGF-IRβ expression. Blots were stripped and used for analysis of β-actin as a control for equal protein loading. (a) Representative Western blot of phospho-IGF-IR or total IGF-IRβ after 22 or 48 hrs. (b, c) Quantitation of total band density is expressed as percent of 22- hr control levels. * p < 0.001 vs CTRL. ** p < 0.05 vs 48 HR IGF-I. (Reprinted from Ness and Wood 2002 with permission from Elsevier).)

IGF-I phophorylation of Akt does not lead to suppression of Bim; rather, activation of Akt suppresses translocation of Bax to the mitochondria through another, as yet unidentified, mechanism (Ness et al. 2004).

The evidence for the PI3K/Akt or other signaling pathways in mediating IGF effects on proliferation or differentiation is just beginning to be investigated. In adult hippocampal progenitor cells, IGF-I appears to use both the PI3K and MAP Kinase (MAPK) pathways to promote proliferation (Aberg et al. 2003). There is considerable evidence that IGF-I mediates proliferation of many transformed cells through activating the MAPK pathway. However, immortalized and transformed cells have been selected for or have altered their survival requirements, which may include alterations in PI3K/Akt signaling. Thus, it is possible that the potent activation of the PI3K/Akt pathway by IGF-I in neural cells may contribute to its effects in mediating proliferation and differentiation, in addition to its ability to promote survival of these cells. Activation of the PI3K pathway recently has been implicated downstream of the IGF-IR in promoting other functions of IGF-I in the CNS, such as glucose utilization (Bondy and Cheng 2004).

Conclusions

In the present review, we provide evidence that IGF-I promotes brain growth through multiple actions on various neural cell types. While originally thought of primarily as a survival factor for neurons and glia in the CNS, there is evidence now that IGF-I is essential for proliferation of mulitpotent stem/progenitor cells and of oligodendrocyte progenitors. Recent investigations into the mechanisms for IGF-I stimulation of neural progenitor proliferation suggest that it enhances G_1 progression and can shorten the length of G_1, partly by coordinating with other mitogens in regulating cell cycle machinery such as cyclin D1. In addition, IGF-I is likely involved in cell fate decisions and differentiation of some progenitor populations. Finally, there is considerable evidence that IGF-I is critical for development of the oligodendrocyte lineage. The ability of IGF-I to promote survival of neural cells appears dependent on its activation of the PI3K/Akt pathway; however, the downstream targets of this pathway in blocking apoptosis may differ between neurons and glia. The pathways responsible for IGF-I actions on proliferation and differentiation have not been clearly elucidated in neural cells. The challenge for future investigations will be to define the signaling pathways and critical downstream targets of IGF-I actions in cell cycle, survival and differentiation pathways in neural cells and to determine how these targets lead to development and growth of the brain.

References

Aberg MA, Aberg ND, Palmer TD, Alborn AM, Carlsson-Skwirut C, Bang P, Rosengren LE, Olsson T, Gage FH, Eriksson PS (2003) IGF-I has a direct proliferative effect in adult hippocampal progenitor cells. Mol Cell Neurosci 24:23-40

Armstrong R, Dorn H, Kufta C, Friedman E, Dubois-Dalcq M (1992) Pre-oligodendrocytes from adult human CNS. J Neurosci 12:1538-1547

Arsenijevic Y, Weiss S (1998) Insulin-like growth factor-I is a differentiation factor for post-mitotic CNS stem cell-derived neuronal precursors: distinct actions from those of brain-derived neurotrophic factor. J Neurosci 18:2118-2128

Arsenijevic Y, Weiss S, Schneider B, Aebischer P (2001) Insulin-like growth factor-I is necessary for neural stem cell proliferation and demonstrates distinct actions of epidermal growth factor and fibroblast growth factor-2. J Neurosci 21:7194-7202

Back S, Gan X, Li Y, Rosenberg P, Volpe J (1998) Maturation-dependant vulnerability of oligodendrocytes to oxidative stress-induced death caused by glutathione depletion. J Neurosci 18:6241-6253

Baker NL, Carlo Russo V, Bernard O, D'Ercole AJ, Werther GA (1999) Interactions between bcl-2 and the IGF system control apoptosis in the developing mouse brain. Brain Res Dev Brain Res 118:109-118

Baron-Van Evercooren A, Olichon-Berthe C, Kowalski A, Visciano G, Van Obberghen E (1991) Expression of IGF-I and insulin receptor genes in the rat central nervous system: a developmental, regional, and cellular analysis. J Neurosci Res 28:244-253

Barres B, Hart I, Coles H, Burne J, Voyvodic J, Richardson W, Raff M (1992a) Cell death and control of cell survival in the oligodendrocyte lineage. Cell 70:31-46

Barres BA, Hart IK, Coles HS, Burne JF, Voyvodic JT, Richardson WD, Raff MC (1992b) Cell death in the oligodendrocyte lineage. J Neurobiol 23:1221-1230

Barres B, Schmid R, Sendtner M, Raff M (1993) Multiple extracellular signals are required for long-term oligodendrocyte survival. Development 118:283-295

Bartlett WP, Li XS, Williams M (1992) Expression of IGF-1 mRNA in the murine subventricular zone during postnatal development. Brain Res Mol Brain Res 12:285-291

Beck K, Powell-Braxton L, Widmer H, Valverde J, Hefti F (1995) Igf1 gene disruption results in reduced brain size, CNS hypomyelination, and loss of hippocampal granule and striatal parvalbumin-containing neurons. Neuron 14:717-730

Behringer RR, Lewin TM, Quaife CJ, Palmiter RD, Brinster RL, D'Ercole AJ (1990) Expression of insulin-like growth factor I stimulates normal somatic growth in growth hormone-deficient transgenic mice. Endocrinology 127:1033-1040

Bondy CA, Cheng CM (2004) Signaling by insulin-like growth factor 1 in brain. Eur J Pharmacol 490:25-31

Bondy C, Werner H, Roberts J, CT, LeRoith D (1990) Cellular pattern of insulin-like growth factor-I (IGF-I) and type I IGF receptor gene expression in early organogenesis: comparison with IGF-II gene expression. Mol Endocrinol 4:1386-1398

Bondy C, Werner H, Roberts J, CT, LeRoith D (1992) Cellular pattern of type I insulin-like growth factor receptor gene expression during maturation of the rat brain: comparison with insulin-like growth factors-I and II. Neuroscience 46:904-923

Bottenstein J, Skaper S, Baron S, Sato G (1980) Selective survival of neurons from chick embryo sensory ganglionic dissociates utilizing serum-free supplemented medium. Exp Cell Res 125:183-190

Brooker GJ, Kalloniatis M, Russo VC, Murphy M, Werther GA, Bartlett PF (2000) Endogenous IGF-1 regulates the neuronal differentiation of adult stem cells. J Neurosci Res 59:332-341

Carson M, Behringer R, Brinster R, McMorris F (1993) Insulin-like growth factor I increases brain growth and central nervous system myelination in transgenic mice. Neuron 10:729-740

Cheng B, Mattson MP (1991) NGF and bFGF protect rat hippocampal and human cortical neurons against hypoglycemic damage by stabilizing calcium homeostasis. Neuron 7:1031-1041.

Cheng B, Mattson MP (1992) IGF-I and IGF-II protect cultured hippocampal and septal neurons against calcium-mediated hypoglycemic damage. J Neurosci 12:1558-1566

Cheng B, Mattson MP (1994) NT-3 and BDNF protect CNS neurons against metabolic/excitotoxic insults. Brain Res 640:56-67.

Cheng B, McMahon DG, Mattson MP (1993) Modulation of calcium current, intracellular calcium levels and cell survival by glucose deprivation and growth factors in hippocampal neurons. Brain Res 607:275-285.

Cheng C, Joncas G, Reinhardt R, Farrer R, Quarles R, Janssen J, McDonald M, Crawley J, Powell-Braxton L, Bondy C (1998) Biochemical and morphometric analyses show that myelination in the insulin-like growth factor 1 null brain is proportionate to its neuronal composition. J Neurosci 18:5673-5681

Chrysis D, Calikoglu AS, Ye P, D'Ercole AJ (2001) Insulin-like growth factor-I overexpression attenuates cerebellar apoptosis by altering the expression of Bcl family proteins in a developmentally specific manner. J Neurosci 21:1481-1489

D'Ercole AJ, Ye P, Calikoglu AS, Gutierrez-Ospina G (1996) The role of the insulin-like growth factors in the central nervous system. Mol Neurobiol 13:227-255

D'Ercole AJ, Ye P, O'Kusky JR (2002) Mutant mouse models of insulin-like growth factor actions in the central nervous system. Neuropeptides 36:209-220

DiCicco-Bloom E, Black IB (1988) Insulin growth factors regulate the mitotic cycle in cultured rat sympathetic neuroblasts. Proc Natl Acad Sci USA 85:4066-4070

Drago J, Murphy M, Carroll SM, Harvey RP, Bartlett PF (1991) Fibroblast growth factor-mediated proliferation of central nervous system precursors depends on endogenous production of insulin-like growth factor I. Proc Natl Acad Sci USA 88:2199-2203

Flores AI, Mallon BS, Matsui T, Ogawa W, Rosenzweig A, Okamoto T, Macklin WB (2000) Akt-mediated survival of oligodendrocytes induced by neuregulins. J Neurosci 20:7622-7630

Frank L, Ventimiglia R, Anderson K, Lindsay RM, Rudge JS (1996) BDNF down-regulates neu-rotrophin responsiveness, TrkB protein and TrkB mRNA levels in cultured rat hippocam-pal neurons. Eur J Neurosci 8:1220-1230.

Frederick TJ, Wood TL (2004) IGF-I and FGF-2 coordinately enhance cyclin D1 and cyclin E-cdk2 association and activity to promote G(1) progression in oligodendrocyte progenitor cells. Mol Cell Neurosci 25:480-492

Gluckman P, Klempt N, Guan J, Mallard C, Sirimanne E, Dragunow M, Klempt M, Singh K, Williams C, Nikolics K (1992) A role for IGF-1 in the rescue of CNS neurons following hy-poxic-ischemic injury. Biochem Biophys Res Commun 182:593-599

Guan J, Williams C, Gunning M, Mallard C, Gluckman P (1993) The effects of IGF-1 treatment after hypoxic-ischemic brain injury in adult rats. J Cereb Blood Flow Metab 13:609-616

Hodge RD, D'Ercole AJ, O'Kusky JR (2004) Insulin-like growth factor-I accelerates the cell cycle by decreasing g1 phase length and increases cell cycle reentry in the embryonic cere-bral cortex. J Neurosci 24:10201-10210

Hsieh J, Aimone JB, Kaspar BK, Kuwabara T, Nakashima K, Gage FH (2004) IGF-I instructs multipotent adult neural progenitor cells to become oligodendrocytes. J Cell Biol 164:111-122

Jiang F, Frederick TJ, Wood TL (2001) IGF-I synergizes with FGF-2 to stimulate oligodendro-cyte progenitor entry into the cell cycle. Dev Biol 232:414-423

Kermer P, Klocker N, Labes M, Bahr M (2000) Insulin-like growth factor-I protects axoto-mized rat retinal ganglion cells from secondary death via PI3-K-dependent Akt phosphor-ylation and inhibition of caspase-3 In vivo. J Neurosci 20:2-8

Knusel B, Gao H, Okazaki T, Yoshida T, Mori N, Hefti F, Kaplan DR (1997) Ligand-induced down-regulation of Trk messenger RNA, protein and tyrosine phosphorylation in rat cor-tical neurons. Neuroscience 78:851-862.

Lee WH, Wang GM, Seaman LB, Vannucci SJ (1996) Coordinate IGF-I and IGFBP5 gene ex-pression in perinatal rat brain after hypoxia-ischemia. J Cereb Blood Flow Metab 16:227-236

Leinninger GM, Backus C, Uhler MD, Lentz SI, Feldman EL (2004) Phosphatidylinositol 3-ki-nase and Akt effectors mediate insulin-like growth factor-I neuroprotection in dorsal root ganglia neurons. FASEB J 18:1544-1546

Lenoir D, Honegger P (1983) Insulin-like growth factor I (IGF I) stimulates DNA synthesis in fetal rat brain cell cultures. Brain Res 283:205-213

LeRoith D, Werner H, Beitner-Johnson D, Roberts CJ (1995) Molecular and cellular aspects of the insulin-like growth factor I receptor. Endocrine Rev 16:143-163

Linseman D, Phelps R, Bouchard R, Le S, Laessig T, McClure M, Heidenreich K (2002) Insulin-like growth factor-I blocks Bcl-2 interacting mediator of cell death (Bim) induction and intrinsic death signaling in cerebellar granule neurons. J Neurosci 22:9287-9297

Liu XF, Fawcett JR, Thorne RG, DeFor TA, Frey WH, 2nd (2001a) Intranasal administration of insulin-like growth factor-I bypasses the blood-brain barrier and protects against focal cerebral ischemic damage. J Neurol Sci 187:91-97

Liu XF, Fawcett JR, Thorne RG, Frey WH, 2nd (2001b) Non-invasive intranasal insulin-like growth factor-I reduces infarct volume and improves neurologic function in rats following middle cerebral artery occlusion. Neurosci Lett 308:91-94

Mason J, Ye P, Suzuki K, D'Ercole A, Matsushima G (2000) Insulin-like growth factor-1 inhib-its mature oligodendrocyte apoptosis during primary demyelination. J Neurosci 20:5703-5708

Mathews LS, Hammer RE, Behringer RR, D'Ercole AJ, Bell GI, Brinster RL, Palmiter RD (1988) Growth enhancement of transgenic mice expressing human insulin-like growth factor I. Endocrinology 123:2827-2833

Matsuzaki H, Tamatani M, Mitsuda N, Namikawa K, Kiyama H, Miyake S, Tohyama M (1999) Activation of Akt kinase inhibits apoptosis and changes in Bcl-2 and Bax expression induced by nitric oxide in primary hippocampal neurons. J Neurochem 73:2037-2046.

Mattson MP, Cheng B (1993) Growth factors protect neurons against excitotoxic/ischemic damage by stabilizing calcium homeostasis. Stroke 24:I136-140; discussion I144-135.

Mattson MP, Zhang Y, Bose S (1993) Growth factors prevent mitochondrial dysfunction, loss of calcium homeostasis, and cell injury, but not ATP depletion in hippocampal neurons deprived of glucose. Exp Neurol 121:1-13.

McCarthy K, de Vellis J (1980) Preparation of separate astroglial and oligodendroglial cell cultures from rat cerebral tissue. J Cell Biol 85:890-902

McDonald JW, Levine JM, Qu Y (1998) Multiple classes of the oligodendrocyte lineage are highly vulnerable to excitotoxicity. NeuroReport 9:2757-2762

McMorris FA, Dubois-Dalcq M (1988) Insulin-like growth factor I promotes cell proliferation and oligodendroglial commitment in rat glial progenitor cells developing in vitro. J Neurosci Res 21:199-209

McMorris F, McKinnon R (1996) Regulation of oligodendrocyte development and CNS myelination by growth factors: Prospects for therapy of demyelinating disease. Brain Pathol 6:313-329

McMorris F, Mozell R, Carson M, Shinar Y, Meyer R, Marchetti N (1993) Regulation of oligodendrocyte development and central nervous system myelination by insulin-like growth factors. Ann NY Acad Sci 692:321-334

McMorris F, Smith T, DeSalvo S, Furlanetto R (1986) Insulin-like growth factor I/somatomedin C: A potent inducer of oligodendrocyte development. Proc Natl Acad Sci USA 83:822-826

Mozell RL, McMorris FA (1991) Insulin-like growth factor I stimulates oligodendrocyte development and myelination in rat brain aggregate cultures. J Neurosci Res 30:382-390

Ness JK, Wood TL (2002) Insulin-like growth factor I, but not neurotrophin-3, sustains Akt activation and provides long-term protection of immature oligodendrocytes from glutamate-mediated apoptosis. Mol Cell Neurosci 20:476-488

Ness JK, Mitchell NE, Wood TL (2002) IGF-I and NT-3 signaling pathways in developing oligodendrocytes: differential regulation and activation of receptors and the downstream effector Akt. Dev Neurosci 24:437-445

Ness JK, Scaduto RC, Jr., Wood TL (2004) IGF-I prevents glutamate-mediated bax translocation and cytochrome C release in O4+ oligodendrocyte progenitors. Glia 46:183-194

Palmiter RD, Brinster RL, Hammer RE, Trumbauer ME, Rosenfeld MG, Birnberg NC, Evans RM (1982) Dramatic growth of mice that develop from eggs microinjected with metallothionein-growth hormone fusion genes. Nature 300:611-615

Palmiter RD, Norstedt G, Gelinas RE, Hammer RE, Brinster RL (1983) Metallothionein-human GH fusion genes stimulate growth of mice. Science 222:809-814

Parrizas M, LeRoith D (1997) Insulin-like growth factor-1 inhibition of apoptosis is associated with increased expression of the bcl-xL gene product. Endocrinology 138:1355-1358

Popken GJ, Hodge RD, Ye P, Zhang J, Ng W, O'Kusky JR, D'Ercole AJ (2004) In vivo effects of insulin-like growth factor-I (IGF-I) on prenatal and early postnatal development of the central nervous system. Eur J Neurosci 19:2056-2068

Roth GA, Spada V, Hamill K, Bornstein MB (1995) Insulin-like growth factor I increases myelination and inhibits demyelination in cultured organotypic nerve tissue. Brain Res Dev Brain Res 88:102-108

Russell JW, Feldman EL (1999) Insulin-like growth factor-I prevents apoptosis in sympathetic neurons exposed to high glucose. Horm Metab Res 31:90-96

Russell JW, Windebank AJ, Schenone A, Feldman EL (1998) Insulin-like growth factor-I prevents apoptosis in neurons after nerve growth factor withdrawal. J Neurobiol 36:455-467

Satoh J, Kim SU (1994) Proliferation and differentiation of fetal human oligodendrocytes in culture. J Neurosci Res 39:260-272

Shinar Y, McMorris F (1995) Developing oligodendroglia express mRNA for insulin-like growth factor-I, a regulator of oligodendrocyte development. J Neurosci Res 42:516-527

Tamatani M, Ogawa S, Tohyama M (1998) Roles of Bcl-2 and caspase in hypoxia-induced neuronal cell death: a possible neuroprotective mechanism of peptide growth factors. Mol Brain Res 58:27-39

Vaccarino FM, Schwartz ML, Raballo R, Nilsen J, Rhee J, Zhou M, Doetschman T, Coffin JD, Wyland JJ, Hung YT (1999) Changes in cerebral cortex size are governed by fibroblast growth factor during embryogenesis. Nature Neurosci 2:246-253

Vemuri G, McMorris F (1996) Oligodendrocytes and their precursors require phosphatidylinositol 3-kinase signaling for survival. Development 122:2529-2537

Vicario-Abejon C, Yusta-Boyo MJ, Fernandez-Moreno C, de Pablo F (2003) Locally born olfactory bulb stem cells proliferate in response to insulin-related factors and require endogenous insulin-like growth factor-I for differentiation into neurons and glia. J Neurosci 23:895-906

Vincent AM, Feldman EL (2002) Control of cell survival by IGF signaling pathways. Growth Horm IGF Res 12:193-197

Vincent AM, Mobley BC, Hiller A, Feldman EL (2004) IGF-I prevents glutamate-induced motor neuron programmed cell death. Neurobiol Dis 16:407-416

Wilkins A, Chandran S, Compston A (2001) A role for oligodendrocyte-derived IGF-1 in trophic support of cortical neurons. Glia 36:48-57

Wood T (1995) Gene-targeting and transgenic approaches to IGF and IGF binding protein function. Am J Physiol 269 (Endocrinol. Metab. 32):E613-E622

Ye P, D'Ercole AJ (1999) Insulin-like growth factor I protects oligodendrocytes from tumor necrosis factor-alpha-induced injury. Endocrinology 140:3063-3072

Ye P, Carson J, D'Ercole AJ (1995) In vivo actions of insulin-like growth factor-I (IGF-I) on brain myelination: studies of IGF-I and IGF binding protein (IGFBP-1) transgenic mice. J Neurosci 15:7344-7356

Ye P, Xing Y, Dai Z, D'Ercole AJ (1996) In vivo actions of insulin-like growth factor-I (IGF-I) on cerebellum development in transgenic mice: evidence that IGF-I increases proliferation of granule cell progenitors. Brain Res Dev Brain Res 95:44-54

Ye P, Li L, Richards G, DiAugustine RP, D'Ercole AJ (2002) Myelination is altered in insulin-like growth factor-I null mutant mice. J Neurosci 22:6041-6051

Zhang Y-Z, Moheban D, Conway B, Bahattacharyya A, Segal R (2000) Cell surface Trk receptors mediate NGF-induced survival while internalized receptors regulate NGF-induced differentiation. J Neurosci 20:5671-5678

Zheng WH, Quirion R (2004) Comparative signaling pathways of insulin-like growth factor-1 and brain-derived neurotrophic factor in hippocampal neurons and the role of the PI3 kinase pathway in cell survival. J Neurochem 89:844-852

IGF-I Deficiency: Lessons from Human Mutations

M.O. Savage[1], C. Camacho-Hübner[1], M.J. Walenkamp[2], L.A. Metherell[1],
A. David[1], L.A. Pereira[2], A. Denley[3], A.J.L. Clark[1] and J.M. Wit[2]

Summary

IGF-I deficiency may be caused by defects in growth hormone (GH) secretion or action. This chapter will focus on genetic mutations causing primary defects of IGF-I synthesis or disturbance of the GH-IGF-I axis resulting in GH insensitivity (GHI). Two patients with mutations of the IGF-I gene have been described. They have several features in common: intra-uterine growth retardation (IUGR), microcephaly, mental retardation, deafness, growth failure and variable insulin resistance. Mutations of the GH receptor (GHR) or downstream signaling pathway or of peptides essential for the formation of the ternary complex also cause IGF-I deficiency, resulting in some disturbance of linear growth. The phenotypic and endocrine features of these mutations causing GHI will also be discussed.

Introduction

The application of molecular biology to the endocrine system has made major contributions to the understanding of basic physiological mechanisms. This is particularly true of defects of the GH-IGF-I axis, where the pathophysiology of certain syndromes associated with growth failure has been clarified by progress in mutation analysis. Although these disorders are usually rare and show striking variation of phenotype, human mutations have demonstrated the key roles played by certain proteins essential for human growth. Predominant amongst these is IGF-I. This chapter will review two basic disorders of IGF-I physiology, First, we discuss primary IGF-I deficiency due to defects in the IGF-I gene itself, and second, defects of GH action directly contributing to the production or stability of circulating IGF-I.

[1] Department of Endocrinology, Barts and the Royal School of Medicine and Dentitry, London, UK
[2] Department of Endocrinology and Metabolic Diseases, Leiden University Medical Centre, Leiden, The Netherlands
[3] Department of Molecular and Biomedical Science, University of Adelaide, Australia

Carel et al.
Deciphering Growth
© Springer-Verlag Berlin Heidelberg

IGF-I gene mutations

Clinical phenotypes

Two patients with mutations of the IGF-I gene have been described. They will be described individually.

Patient 1

In 1996, the first patient with severe growth failure due to primary IGF-I deficiency caused by a partial homozygous deletion of the IGF-I gene was reported (Woods et al. 1996a; Camacho-Hübner et al. 2002). The most important clinical features were severe intra-uterine growth retardation (IUGR) and post-natal growth failure. The patient, a white Caucasian male, was born by caesarean section at 37 weeks gestation with a birth weight of 1.4 kg (-3.9 SD below the mean), birth length of 37.8 cm (-5.4 SD) and a head circumference of 27 cm (-4.9 SD). Growth failure continued throughout childhood. He had severe bilateral sensorineural deafness and moderate developmental delay and hyperactivity. At age 15.2 years, he was referred to our hospital with a possible diagnosis of GH insensitivity (GHI) syndrome, GH deficiency having been excluded earlier.

Clinical evaluation showed that his height was 119.1 cm, (-6.7 SD), weight 23 kg (-6.5 SD) and BMI 16.2 kg/m² (-1.9 SD). He had dysmorphic features, not seen in patients with Laron syndrome, consisting of micrognathia, a low hairline, ptosis and severe microcephaly. He was in early puberty with testicular volumes of 4 ml bilaterally. In addition, he had bilateral hearing loss and mild myopia. RhIGF-I therapy in a dose of 80 µg/kg/day increased his growth rate from <3 cm/yr to 6.5 cm/yr. His final height is 130.2 cm. He progressed normally through puberty.

Patient 2

This patient, also a white Caucasian male, was born as the first of five children of a consanguineous marriage (Walenkamp et al. 2005). Birth weight was 1420 g (-3.9 SD) and length 39 cm (-4.3 SD). Postnatally, he had persistent progressive growth failure with normal proportions, retarded skeletal maturation, microcephaly, deaf-mutism and severe mental retardation. At age 15 years, he was institutionalised because of severe mental retardation (IQ<40). Pubic hair and testicular growth occurred at the age of nearly 20 years. At age 55 years, he came to medical attention again because of a request for genetic counselling by one of his healthy brothers. Abdominal fat mass was increased and was associated with several dysmorphic features, including deep-set eyes, flat occiput, a columella extending beyond the alae nasi, and striking micrognathia. Extremities showed broad end phalanges and convex nails. There was severe bilateral hearing loss confirmed by absent brainstem evoked potentials. Testicular volume was 7 ml (left) and 1 ml (right). His youngest brother was born at term with a birth weight of 1900 g (-4.5 SD; van Gemund et al. 1969), with a clinical phenotype closely resembling the index case. Photographs of the two patients are shown in Figure 1.

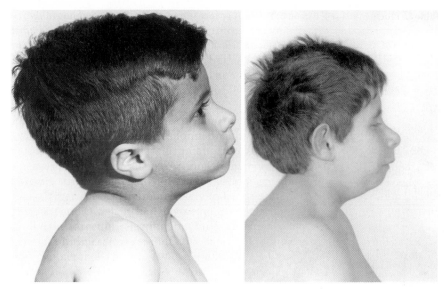

Fig. 1. Phenotypic similarity in the two patients with IGF-I gene mutations. Patient 1 (right), Patient 2 (left)

Biochemical features

Patient 1

The results of the initial biochemical assessment of the GH-IGF-I axis (Camacho-Hübner et al.1999) are summarised in Table 1. In addition to his GH response during an insulin tolerance test with a GH peak of 61 ng/ml, he had serum GH concentrations, measured every 20 minutes from 20.00 h to 08.00 h, that ranged from 2.2 ng/ml to 171 ng/ml. The patient also had marked hyperinsulinaemia with fasting euglycaemia. Insulin sensitivity was assessed using the modified Bergman's protocol (Woods et al. 2000) and shown to be decreased. On rhIGF-I therapy, in a dose of 40 µg/kg/day increasing to 80 µg/kg/day, insulin sensitivity significantly improved (Woods et al. 2000; Fig. 2). The changes to the constituent peptides of the GH-IGF-I axis during rhIGF-I therapy (Camacho-Hübner et al. 1999) are summarised in Table 1.

Patient 2

The results of the biochemical analysis are shown in Table 2. Maximum GH concentration was in the upper normal range. Serum IGF-I was markedly elevated. Serum concentrations of IGFBP-2, -3, -4, -5 and -6 were within the normal range, whereas acid-labile subunit (ALS) was increased.

Table 1. Biochemical assessment of Patient 1 at baseline and during IGF-I therapy.

Peptide	Basal	rhIGF-I 40 µg/kg/d	rhIGF-I 80 µg/kg/d
IGF-II (ng/ml)	1044	888	756
ALS (mg/L)	46.3	46	30
IGFBP-3 (mg/L)	5.8	6.9	4.7
IGFBP-2 (ng/ml)	73	112	194
Insulin (mU/L)	27.3	21	12
IGFBP-1 (ng/ml)	4.7	5.5	26.8

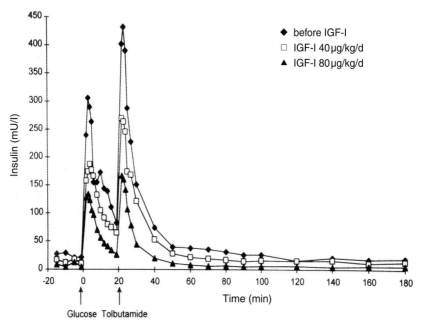

Fig. 2. Decrease in insulin secretion, using the Bergman FSIVGTT, during rhIGF-I therapy in Patient 1 with IGF-I gene deletion.

Table 2. Evaluation of the GH-IGF-I axis in Patient 2.

GH	206 ng/ml (300 mU/L)	(ITT, age 22 yr)
IGF-I	79 nmol/L	(+ 7.3 SDS)
IGFBP-3	66 nmol/L	(+ 0.1 SDS)
IGF-II	60.8 nmol/L	(0.5 SDS)
Insulin	13 mU/L	increased
Glucose	6.2 mmol/L	

Molecular studies

Patient 1
Molecular studies demonstrated deletion of exons 4 and 5 of the IGF-I gene. Skin fibroblasts from the patient had a reverse transcriptase PCR product that was 182 bp shorter than in normal subjects. Sequencing studies showed that exon 3 continued directly into exon 6, confirming the deletion of exons 4 and 5 (Woods et al. 1996).

Patient 2
IGF-I cDNA was isolated by RT-PCR from fibroblasts. Sequence analysis identified a homozygous G>A nucleotide substitution at position 274, changing valine at position 44 of the mature IGF-I protein to methionine (V44M). The same nucleotide substitution was also present in the genomic DNA but not in a control panel of 87 individuals. Functional studies demonstrated that V44M IGF-I exhibits a 90-fold decrease in type-1 IGF receptor (IGF-1R) binding, compared with wild-type human IGF-I, and only poorly stimulates autophosphorylation of the IGF-1R. Consequently, Val44 has been identified as an essential residue involved in the IGF-IGF-1R interaction (Denley et al. 2004).

A family study was undertaken to genotype 24 relatives of the index patient. Nine carried the heterozygous V44M IGF-I mutation. Their birth weights, head circumferences and heights were lower than in the non-carriers (Walenkamp et al. 2005). IGF-I levels and fasting insulin levels were higher and IGFBP-1 levels were lower than in the non-carriers (Walenkamp et al. 2005).

Other mutations in the GH-IGF-I axis causing IGF-I deficiency and GH insensitivity

Spectrum of GH insensitivity:
Laron syndrome, atypical GHI syndrome and idiopathic short stature

GHI is classified into primary or genetic GHI and secondary or acquired GHI (Laron et al. 1993). Laron reported the first case of primary GHI (Laron 2004).

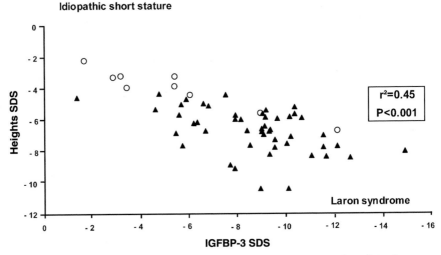

Fig. 3. Phenotypic variation from Laron syndrome (triangles) to idiopathic short stature (circles) in 59 children with GH insensitivity and correlation between height SDS and IGFBP-3 SDS.

Its aetiology consists of genetic defects in the GH receptor (GHR). Most classical GHI patients have high circulating GH and low serum levels of GHBP, IGF-1 and IGFBP-3, but some have normal or even high GHBP levels (Woods et al. 1997). Some patients with primary GHI have a less severe disorder, known as atypical GHI, that is associated with a normal facial appearance and less abnormal growth and biochemical features (Fig. 3). As these patients do not have the phenotype of Laron syndrome, they may be categorized as having idiopathic short stature (Burren et al. 2001).

GH receptor (GHR) mutations

The cloning and characterization of the human GHR (Leung et al. 1987) have permitted the understanding of the pathophysiology of GHI. The ease with which it is now possible to amplify the coding exons of GHR by PCR and sequence the amplification products has led to the discovery of more than 60 mutations in the GHR gene of GHI patients. They are almost all recessively inherited, either in homozygous or compound heterozygous forms, and range from exon deletions to a variety of point mutations, including missense, nonsense, splice, and frameshift mutations (Woods et al. 1997; Baumann 2002).

Nearly all reported molecular defects in GHR occur in the region encoding the extracellular domain of the receptor. Patients with such mutations have absent or extremely low GHBP levels and the typical facial features of Laron syndrome (Woods et al. 1997). More than 50 mutations in this domain have been reported [(Baumann 2002), including deletion of a large portion of the gene and small deletions resulting in frameshifts and therefore premature stop codons

More common are point mutations that result in a premature stop codon (non-sense), or altered amino acid (missense) and nucleotide substitutions that result in activation of a cryptic splice site or creation of a new one.

Mutations in GHBP-positive GHI syndrome

Duquesnoy et al. (1994) reported several patients with a D152H mutation and normal GHBP levels. We have also described a classical GHI syndrome patient with a homozygous point mutation (IVS8ds+1 G→C) in the splice donor site of intron 8, resulting in the skipping of exon 8. This was the first description of an intracellular mutation and we predicted that the mutant GHR would have been released from cells and measured as GHBP, lacking the ability to bind to the cell surface (Woods et al. 1996b).

Atypical GHI syndrome

In 1997, Ayling et al. provided new insights into the molecular defects in GHI, describing the first heterozygous mutation with a dominant negative effect. The authors reported a mutation (IVS8as-1 G→C) in the acceptor splice site of intron 8, resulting in the skipping of exon 9 and the production of a truncated GHR. The mutant GHR formed heterodimers with the wild type GHR and exerted a dominant negative effect on the normal protein (Ayling et al. 1997). A second mutation (IVS9ds+1 G→A) leading to the same consequence was described by Iida et al. (1998). Both of these patients have normal GHBP and normal facial appearance without the characteristic features of Laron syndrome (Burren et al. 2001). We described a similar phenotype in four patients from a consanguineous pedigree with a mutation causing the insertion of a pseudoexon between exons 6 and 7. This 108 bp insertion caused the addition, in-frame, of 36 amino acids between codons 206 and 207 (Metherell et al. 2001).

Idiopathic short stature (ISS)

Although more then 60 molecular defects in the GHR have been described, there are still short stature patients with normal GH secretion, in whom the cause of growth failure has not yet been completely explained. In 1995, Goddard et al. studied a group of ISS patients with low serum concentrations of GHBP, suggestive of partial GHI, and described heterozygous mutations of the GHR (Goddard et al. 1995, 1997). Several other mutations of the GHR have been described in patients with short stature and GHI without features of Laron syndrome (Blair and Savage, 2002; Fig. 4).

STAT5b and ALS gene defects

Two studies in the past have shown the absence of GH-induced tyrosine phosphorylation of the STAT protein in patients with ISS and no mutations in the GHR gene, but the authors were not able to identify mutations in these patients. Only recently did the first report of a molecular defect in the GH signalling cas-

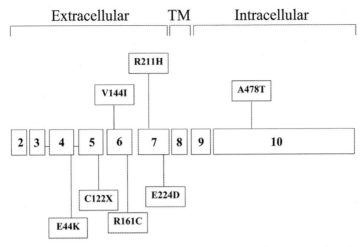

Fig. 4. Growth hormone receptor (GHR) structure and GHR mutations in patients with idiopathic short stature (ISS). Exons from 2 to 7 encode the extracellular domain, exon 8 the transmembrane domain (TM) and exons 9 and 10 the intracellular domain of GHR. Identified mutations in ISS patients with low GHBP levels are indicated in boxes.

cade in an ISS patient appear. Kofoed et al. (2003) reported a homozygous mutation in exon 15 of the STAT5b gene and demonstrated that the mutant protein could not be stimulated by the GHR, therefore failing to activate gene transcription. This child had features of classical GHI and also had immunodeficiency consistent with a non-functional STAT5b.

A second interesting defect in the GH-IGF-I axis was recently reported by Domene et al. (2004), who described the first mutation of the ALS gene. This homozygous mutation prevented the stability of the circulating ternary complex, leading to a severe deficiency of IGF-I. This patient, a male aged 16 years, had short stature but not the extreme growth failure seen in Laron syndrome. Also, his facial features were normal. It is possible that the relative mildness of this phenotype is related to the fact that paracrine IGF-I production may be normal compared to circulating IGF-I levels, which are markedly reduced.

Conclusions

The description of the second case of human mutation of the IGF-I gene has strengthened the conclusions that IGF-I is essential for fetal and post-natal growth and apparently for brain growth, intellectual development and normal hearing. The insulin resistance demonstrated by one of the two patients requires further study and may theoretically be attributable to either deficiency of IGF-I per se or to the direct metabolic effects of hypersecretion of GH, which both patients demonstrated.

Table 3. Variation of height SDS and facial appearance in patients with six mutations in the GH-IGF-I axis causing IGF-I deficiency.

No	Mutation	Age (yr)	IGF-I (ng/l)	Height SDS	Appearance
1	Homozygous Ex 4 *GHR*	7	< 20	- 8.4	Laron syndrome
2	Homozygous Ex 8 *GHR*	4	< 20	- 5.6	Laron syndrome
3	Homozygous Intr 6-7 *GHR*	10	21	- 4.4	Normal
4	Homozygous Intr 8-9 *GHR*	11	43	- 4.4	Normal
5	Homozygous *STAT 5b*	12	31-55	- 7.5	Laron syndrome
6	Homozygous *ALS*	14	31	- 2.2	Normal

The phenotypes of patients with IGF-I deficiency due to mutations causing GHI are variable. Height SDS values in a range of human mutations are shown in Table 3. It appears that the classical facial appearance of Laron syndrome, also present in genetic GH deficiency, may only occur when there is severe IGF-I deficiency. As we have seen, GHI due to a number of different mutations may occur in a milder form and be associated with a normal facial appearance and with growth failure that is less severe than typical in classical Laron syndrome. As further human mutations are identified, the pivotal role of IGF-I production in regulating linear growth and other human phenotypic features will become increasingly understood.

References

Ayling RM, Ross R, Towner P, Von Laue S, Finidori J, Moutoussamy S, Buchanan CR, Clayton PE, Norman MR (1997) A dominant-negative mutation of the growth hormone receptor causes familial short stature. Nature Genet 16:13-14.

Baumann G (2002) Genetic characterization of growth hormone deficiency and resistance: implications for treatment with recombinant growth hormone. Am J Pharmacogenomics 2:93-111

Blair JC, Savage MO (2002) The GH-IGF-I axis in children with idiopathic short stature. Trends Endocrinol Metab 13: 325-330

Burren CP, Woods KA, Rose SJ, Tauber M, Price DA, Heinrich U Gilli G, Razzaghy-Azar M, Al-Ashwal A, Crock PA, Rochiccioli P, Yordam N, Ranke MB, Chatelain PG, Preece MA, Rosenfeld RG, Savage MO (2001) Clinical and endocrine characteristics in atypical and classical growth hormone insensitivity syndrome. Horm Res 55:125-130.

Camacho-Hübner C, Woods KA, Miraki-MoudF, Hindmarsh PC, Clark AJ, Hansson Y, Johnston A, Baxter RC, Savage MO (1999) Effects of recombinant human insulin-like growth factor (IGF)-I therapy on the growth hormone (GH)-IGF system of a patient with a partial IGF-I gene deletion. J Clin Endocrinol Metab 84:1611-1616

Camacho-Hübner C, Woods KA, Clark AJL, Savage MO (2002) Insulin-like growth factor (IGF)-I gene deletion. Rev Endoc Metab Dis 3:357-361

Denley A, Wang CW, McNeil KM, Walenkamp MJE, van Duyvenvoorde H, Wit JM, Wallace JC, Norton RS, Karperien M, Forbes BF (2004) Structural and functional characteristics of the Val44Met IGF-I missense mutation: correlation with effects on growth and development. Mol Endocrinol 10.1210

Domene HM, Bengolea SV, Martinez AS, Ropelato MG, Pennisi P, Scaglia P, Heinrich JJ, Jasper HG (2004) Deficiency of the circulating insulin-like growth factor system associated with inactivation of the acid-labile subunit gene. New Eng J Med 350:570-577

Duquesnoy P, Sobrier ML, Duriez B, Dastot F, Buchanan CR, Savage MO Preece MA, Craescu CT, Blouquit Y, Goossens M, Amselem S. (1994) A single amino acid substitution in the exoplasmic domain of the human growth hormone (GH) receptor confers familial GH resistance (Laron syndrome) with positive GH-binding activity by abolishing receptor homodimerization. EMBO J 13:1386-1395.

Goddard AD, Covello R, Luoh SM, Clackson T, Attie KM, Gesundheit N, Rundle AC, Wells JA, Carlsson LM (1995) Mutations of the growth hormone receptor in children with idiopathic short stature. The Growth Hormone Insensitivity Study Group. New Engl J Med 333:1093-1098.

Goddard AD, Dowd P, Chernausek S, Geffner M, Gertner J, Hintz R, Hopwood N, Kaplan S, Plotnick L, Rogol A, Rosenfeld R, Saenger P, Mauras N, Hershkopf R, Angulo M, Attie K (1997) Partial growth-hormone insensitivity: the role of growth-hormone receptor mutations in idiopathic short stature. J Pediatr 131(1 Pt 2):S51-55.

Iida K, Takahashi Y, Kaji H, Nose O, Okimura Y, Abe H, Chihara K (1998) Growth hormone (GH) insensitivity syndrome with high serum GH-binding protein levels caused by a heterozygous splice site mutation of the GH receptor gene producing a lack of intracellular domain. J Clin Endocrinol Metab 83:531-537.

Kofoed EM, Hwa V, Little B, Woods KA, Buckway CK, Tsubaki J, Pratt KL, Bezrodnik L, Jasper H, Tepper A, Heinrich JJ, Rosenfeld RG (2003) Growth hormone insensitivity associated with a STAT5b mutation. New Engl J Med 349:1139-1147

Laron Z (2004). Laron syndrome (primary growth hormone resistance or insensitivity): the personal experience 1958-2003. J Clin Endocrinol Metab 89:1031-1044.

Laron Z, Blum W, Chatelain P, Ranke M, Rosenfeld R, Savage M, Underwood L (1993) Classification of growth hormone insensitivity syndrome. J Pediatr 122:241.

Leung DW, Spencer SA, Cachianes G, Hammonds RG, Collins C, Henzel WJ, Barnard R, Waters MJ, Wood WI (1987) Growth hormone receptor and serum binding protein: purification, cloning and expression. Nature 330:537-543

Metherell LA, Akker SA, Munroe PB, Rose SJ, Caulfield M, Savage MO, Chew SL, Clark AJL (2001) Pseudoexon activation as a novel mechanism for disease resulting in atypical growth-hormone insensitivity. Am J Human Genet 69:641-646.

van Gemund JJ, Laurent de Angulo MS, van Gelderen HH (1969) Familial prenatal dwarfism with elevated serum immuno-reactive growth hormone levels and end-organ unresponsiveness. Maandschr Kindergeneeskd 37:372-382

Walenkamp MJE, Karperien M, Pereira AM, Hilhorst-Hofstee Y, van Doorn J, Chen JW, Mohan S, Denley A, Forbes B, van Duyvenvoorde HA, van Thiel SW, Sluimers CA, Bax JJ, de Laat JAPM, Breuning MB, Romijn JA, Wit JM (2005) Homozygous and heterozygous expression of a novel IGF-I mutation. J Clin Endocrinol Metab 90:2855-2864

Woods KA, Camacho-Hubner C, Savage MO, Clark AJ (1996a) Intrauterine growth retardation and postnatal growth failure associated with deletion of the insulin-like growth factor I gene. New Engl J Med 335:1363-1367

Woods KA, Fraser NC, Postel-Vinay MC, Savage MO, Clark AJ (1996b) A homozygous splice site mutation affecting the intracellular domain of the growth hormone (GH) receptor resulting in Laron syndrome with elevated GH-binding protein. J Clin Endocrinol Metab 81:1686-90.

Woods KA, Dastot F, Preece MA, Clark AJ, Postel-Vinay MC, Chatelain PG, Ranke MB, Rosenfeld RG, Amselem S, Savage MO (1997) Phenotype: genotype relationships in growth hormone insensitivity syndrome. J Clin Endocrinol Metab 82:3529-3535.

Woods KA, Camacho-Hübner C, Bergman RN, Clark AJ, Savage MO (2000) Effect of insulin-like growth factor I (IGF-I) therapy on body composition and insulin resistance in IGF-I gene deletion. J Clin Endocrinol Metab 85:1407-1411

Putting IGF-I Biology into a Clinical Perspective

P.E. Clayton[1], U. Das[1] and A.J. Whatmore[1]

Introduction

IGF-I is a markedly growth hormone (GH)-dependent circulating peptide that is used extensively in the investigation of growth disorders. Within the circulation, IGF-I is bound to its partner binding proteins, with the majority being present in a ternary complex comprising IGF-I, IGFBP-3 and the acid labile subunit (ALS). IGFBP-3 and ALS are also GH-dependent. Measurements of IGF-I and IGFBP-3 are used to confirm the diagnosis of GH deficiency as defined by GH stimulation testing (Report of a Workshop 2000), as initial tests in a short child to determine whether GH testing is indicated, and to establish whether the growth disorder is associated with IGF-I deficiency (Rosenfeld 1997). In more recent years in paediatric practice, IGF-I measurement has been used to monitor GH treatment, as a marker of compliance and as a safety parameter (Juul et al. 2002).

IGF-I is generated from a single copy gene that has two alternative leader exons (1 and 2) and two alternative 3' exons (5 and 6), with the latter also being transcribed as a spliced portion of exon 5 placed between exons 4 and 6 (Fig. 1). The precise mechanisms that determine the transcription of each mRNA are not fully understood, but mRNAs containing exon 2 and exon 5 sequences are considered to be more GH-dependent (Lowe et al. 1988; O'Sullivan et al. 2002). The mature IGF-I sequence is contained within exons 3 and 4. Three prohormones (Ia, Ib and Ic) are generated and are then cleaved by furin to release three carboxy terminal E peptides (Ea, Eb, and Ec). A range of functions has been ascribed to the E peptides; it is possible that, at the pericellular level, these peptides contribute to IGF-I action. The Eb peptide has been shown to induce neurite outgrowth and inhibit anchorage-dependent growth in neuroblastoma cells (Kuo and Chen 2002), whereas the Ec peptide (or muscle growth factor) inhibits terminal differentiation, increases myoblast proliferation and enhances muscle regeneration (Musaro et al. 2004). There is a need to develop techniques to detect these peptides, so that their exact role and their relationship to IGF-I action can be defined.

[1] Academic of Child Health, Division of Human Development & Reproductive Health, The Medical School, University of Manchester, M13 9PT Manchester, England

Carel et al.
Deciphering Growth
© Springer-Verlag Berlin Heidelberg

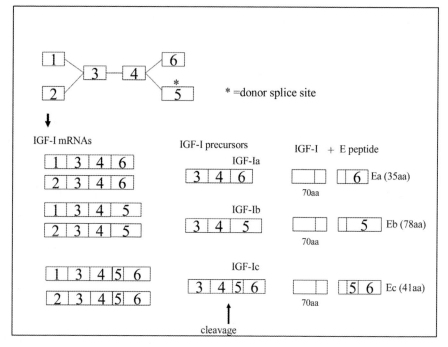

Fig. 1. IGF-I exons, mRNAs, prohormones and E peptides

IGF Parameters for the Diagnosis of GH Deficiency

Severe, congenital GH deficiency is invariably associated with low serum IGF-I levels. In fact, in a study of IGF parameters in a cohort of GH-deficient individuals from Brazil with a homozygous mutation within the GH-releasing hormone receptor gene, IGF-I, IGFBP-3, ALS and IGF-II were all reduced compared to levels in age-matched indigenous controls, with the most significant reduction being seen for IGF-I (Fig. 2; Aguiar-Oliveira et al. 1999).

Over the last 15 years, there have been many reports on the use of IGF-I and IGFBP-3 to support the diagnosis of GH deficiency (GHD). Most studies have assessed the performance of IGF parameters in relation to a diagnosis of GHD based on peak GH levels during provocative tests (Clayton 1999). We recognise that such tests have limitations with a high false positive rate, in which GHD is incorrectly diagnosed. Such studies have indicated a wide range of diagnostic performance but, on average, for IGF-I similar levels of sensitivity (the percentage of abnormal tests in GHD individuals) and specificity (the percentage of normal tests in normal individuals) of ~70% have been found (Table 1). IGFBP-3 measurements have the advantage of a higher specificity but lower sensitivity. It is notable, however, that diagnostic performance for both IGF-I and IGFBP-3 varies markedly between studies, related to assay performance, patient selection and the standard against which the diagnostic performance is compared.

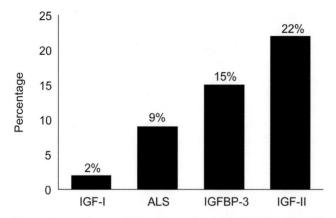

Fig. 2. Impact of severe GHD on serum levels of IGF-I, IGF-II, IGFBP-3 and ALS. Each column represents the percentage value of the mean level in severe GHD individuals with a homozygous GHRH-R mutation compared to the mean value in age-matched controls.

Table 1. Performance of serum IGF-I and IGFBP-3 in the diagnosis of GHD. Data are expressed as mean percentage and range and are based on 13 peer-reviewed publications (Clayton 1999).

	Sensitivity (%)	Specificity (%)
IGF-I	73 (47-100)	73 (47-98)
IGFBP-3	66 (15-97)	85 (57-98)

A recent study of the performance of IGF-I and IGFBP-3 using receiver operator curves in the diagnosis of GHD based on a peak GH of <7 µg/L revealed a sensitivity for IGF-I of 68% and a specificity of 97% at an IGF-I cut-off of -1.65 SDS (Boquete et al. 2003). For IGFBP-3, sensitivity was 60% and specificity was 90% at a cut-off of -1.8 SDS. In view of the difficulties and limitations in defining exactly what the limits for GHD should be, it is unlikely that we will be able to refine the diagnostic performance of these tests any further. They must form part of a multi-faceted approach to the diagnosis of GHD based on history, examination, biochemical status of the GH-IGF axis and hypothalamic-pituitary appearances on MR scanning.

Monitoring IGF-I levels on GH treatment

Monitoring of the child receiving GH treatment has traditionally involved the assessment of growth response and clinical surveillance for side effects. In more recent years, regular measurement of IGF-I has been undertaken by paediatric endocrinologists: in a survey by the European Society of Paediatric Endocrinology, 80% of clinicians reported that this was routine practice (Juul et al. 2002). However, there are very limited data reported on the long-term use of IGF-I monitoring in GH-treated children, and almost nothing on the interpretation of such data or the strategies for GH dose adjustment.

Despite this dearth of information, consensus statements on the management of GH treatment in childhood have advocated that IGF-I and IGFBP-3 should be routinely monitored (Report of a Workshop 2000; Report of a Workshop 2001). The principle that a marker of GH action, such as IGF-I, should be measured is a reasonable one, although the evidence that IGF-I is, in fact, a precise marker is not wholly justified. Nevertheless, in adults with GHD, the dose of GH for replacement is determined by titration against IGF-I levels, aiming to generate a level between the mean and +2 standard deviations (SD) above the mean (i.e., high normal) and to keep the patient side –effect-free (Drake et al. 1998).

Safety Issues

GH has been associated with a wide range of side effects (including arthralgia, fluid retention and benign intracranial hypertension, glucose intolerance, and skeletal problems), although the incidence of significant problems is low. The concerns about side effects are particularly prominent in those non GH-deficient conditions where high doses are used through childhood and adolescence. One potential consequence of GH treatment that has received considerable attention has been the concern about the induction of de novo malignancy or, in cancer survivors, the induction of first tumour recurrence or an increase in the incidence of second tumours. The focus on this potential problem has been heightened by the epidemiological observations that those in the general population with the highest levels of IGF-I have increased risk of the development of a range of malignancies (prostate, breast, colon), with a concomitant low IGFBP-3 level in some instances increasing the significance of the association (Chan et al. 1998; Hankinson et al. 1998; Yu et al. 1999; Petridou et al. 1999). These data are derived from individuals presumably exposed to a life time of relatively high IGF-I (and relatively low IGFBP-3). There are, however, no data on risk of a relatively elevated IGF-I over the growing years. For instance, would this be a particularly sensitive period for exposure to IGF-I?

A recent analysis of cancer incidence in UK recipients of human pituitary GH between 1959 and 1985 revealed an increased incidence of colon and rectal cancer and an increased mortality form colon/rectal cancer and Hodgkin's lymphoma, even when high-risk groups were excluded from analysis (Swerdlow et al. 2002). This cohort of patients was, of course, treated in a very different way than the standard practice of today: the GH was pituitary derived, the doses were given two to three times per week rather than daily, there was no dose adjustment based on size and there was no monitoring of IGF-I. Reassuringly, a more recent survey from the US revealed no increased incidence of tumours in the years after GH (Wyatt 2004).

It is important, however, to maintain high levels of surveillance for any untoward effect of GH, and now that IGF-I data are more readily available, to relate levels to the possible genesis of any adverse events.

Initial Changes in IGF-I and IGFBP-3 during GH Treatment

Many investigators have reported changes in IGF-I during the first year on GH treatment, and the majority have shown that change in IGF-I is a relatively weak correlate of change in growth rate, even in GHD. Recent studies have also reported IGFBP-3 levels and use the ratio of IGF-I to IGFBP-3 as a crude index of bioavailable IGF-I (Ranke et al. 2001). One very important aspect when reporting any monitoring data is the expression of the data, which is best done by the use of SD scores, which requires a well-characterised normative data-set but then allows ready comparison both between individuals and within an individual over time.

We compared IGF-I and IGFBP-3 levels in prepubertal GHD children compared to a non-GHD group [Turner [TS] and Noonan syndrome and small-for-gestational age (SGA; Tillmann et al. 2000)]. A number of observations can be made:

- There is a different pattern of change in IGF-I and IGFBP-3 between GHD and non-GHD children, with the former showing a rapid rise in both parameters and the latter only achieving a significant increase in IGF-I and IGFBP-3 by the end of the first year on GH
- There is a wide range of values for both parameters in both groups, implying a wide range of sensitivity to GH
- Few children were over-treated, with values at or >+2 SD only for IGF-I
- There was considerable variability in (IGF-I - IGFBP-3) sds in the non-GHD group, with no significant change during the first year of GH treatment
- In this small study, change in IGF parameters did not correlate with growth rate. The only significant marker for growth performance was baseline serum leptin level

A similar study, reported by Lanes and Jakubowicz (2002), recognised that, in prepubertal GHD, IGF-I increases more rapidly than IGFBP-3, that IGF-I:IGFBP-3 did not correlate with GH dose and that change in IGF-I did not correlate with change in growth rate.

Cohen et al. (2002) have provided comprehensive IGF-I and IGFBP-3 data over the first two years of treatment with GH in prepubertal GHD children. Three doses of GH were used in the study (normal replacement, 0.025 mg/kg/day; moderate dose, 0.05 mg/kg/day; high dose 0.1 mg/kg/day). Maximum growth performance was achieved with the moderate dose schedule, whereas changes in IGF-I and, to a lesser extent, IGFBP-3 were dose-dependent. Thus high-dose regimens are likely to induce abnormal IGF-I levels but no further improvement in growth rate. This study also demonstrated that the best growth occurred in those with the highest levels of both IGF-I and IGFBP-3, and the reciprocal applied. Thus in GHD, IGF-I needs to be "buffered" by IGFBP-3 to achieve the best growth result. This work identified sexual dimorphism in response to GH, a phenomenon readily recognised in adult GHD but not reported in childhood studies. The dose dependence in change in growth and IGF-I was seen much better in boys than in girls, implying that the latter are more GH-insensitive.

One of the largest studies of IGF monitoring was reported by Ranke et al. (2001), with 156 GHD and 153 non-GHD children (TS, SGA and idiopathic short stature) followed over four years on standard GH regimens. Their data showed the following:

- Tests >95th percentile
1) The incidence of abnormal tests was higher in the non-GHD than in the GHD group
2) The IGF-I:IGFBP-3 ratio was most likely to be abnormally high
3) The number of abnormal tests increased in the pubertal groups

- Tests <5th percentile
1) GHD children were more likely to have low tests than the non-GHD
2) The IGF-I:IGFBP-3 ratio was least likely to be abnormal
3) Low levels were less frequent in puberty

These data were also used to look at whether change in IGF parameters predicted growth response to GH. In the GHD group, all IGF parameters (basal and change over the first three months) correlated with growth rate with the exception of IGF-I:IGFBP-3. In the non-GHD group, only basal IGFBP-3 and change in IGF-I, IGFBP-3 and –2 were significantly correlated to growth performance. This finding implies that the change in the IGF axis in response to GH varies dependent on the diagnosis, and any management strategy must be disease oriented.

Long-term IGF-I and IGFBP-3 Monitoring in TS and SGA

IGF-I levels have been reported in the Dutch TS study, in which children were treated with three different regimens of GH (van Pareren et al. 2003a). The baseline status of IGF-I was low, followed by significant increments to values well above the mean by six months. IGF-I continued to increase over the next two to three years, with dose dependency particularly between the lowest dose regimen and the two higher doses. After three years, the average IGF-I SD scores in the two higher dose groups were consistently >+2. The IGF-I values fell to the mean after treatment in all groups. The significance of maintaining IGF-I levels above the normal range for five to seven years in TS girls is unknown.

Similar data were found in SGA children treated with 0.033 mg/kg/day or 0.067 mg/kg/day (van Pareren et al. 2003b). The children had low IGF-I and IGFBP-3 levels at the start. Peak values took longer to achieve in the low-dose group (four years for IGF-I) than in the high-dose group (three years for IGF-I), but in both groups the mean level was close to or at +2 SD [+1.8 (sd 0.8) versus +2 (sd 1.2)]. At discontinuation, IGF-I remained elevated at +1 and +1.3, respectively, for the two dose groups.

Thus monitoring of IGF-I levels in these groups is of particular importance not only for documentation but also to consider whether adjustment in treatment regimens is required.

Table 2. Baseline data in the three GH-treated groups in a cross-sectional survey of IGF-I and IGFBP-3 levels during GH treatment (Das et al. 2003).

Variable	GHD	TS	Non-GHD
Age at start of GH (yr)	5.2, 0.1-16.9	6.8, 1.7-15.5	9.9, 1.9-13.7
Time on GH (yr)	4.2, 1-13.8	4.1, 0.2-10.7	3.1, 1-12.7
ΔHt sds over year to sample	+ 0.2 (0.2)	+ 0.02 (0.4)	+ 0.2 (0.4)
GH dose (mg/m^2/d) [mg/kg/d]	0.8, 0.3-1.7 [0.028]	1.1, 0.4-1.6 [0.039]	0.9, 0.6-1.5 [0.032]

Cross-sectional Survey of IGF Monitoring

In view of the paucity of IGF monitoring data throughout the GH treatment years, we undertook a cross-sectional study to assess IGF-I and IGFBP-3 in all patients on GH, irrespective of the time on GH (Das et al. 2003). We had a number of aims:
- Establishing the incidence of abnormal tests
- Assessing the number of times a change in GH dose may be indicated
- Examining the relationship between IGF values and growth performance

The children were on average 5 to 10 years of age at the start of GH treatment, and had been treated for a median time of three to four years, with ranges up to 14 years. Standard GH treatment regimens were used (Table 2).

The range of diagnoses in this study is shown in Table 3. There was a weak but significant correlation between change in height SDS in the year up to sampling and the IGF-I SDS (Fig. 3). There were a significant number of tests (26% in total) where a change in GH treatment might be indicated.

In univariate analysis, IGF-I sds was the most significant correlate of growth rate in the GHD group (Table 4). In the TS group, growth rate was inversely correlated with GH dose. This finding probably reflects an iatrogenic effect, where TS girls with the slowest growth would have their GH dose raised. In the non-GHD group, no IGF parameters correlated with growth.

We also examined. in multiple regression analysis, those parameters that influenced growth performance (Table 5). In the GHD group, it is notable that both (IGF-I – IGFBP-3) SDS and IGFBP-3 SDS featured as positive independent predictors, the latter finding being similar to the results of the study of Cohen et al. (2002). Dose was only a significant variable if expressed per m^2 rather than per kg. Background auxological parameters (discrepancy of height from target height and BMI), which feature in first year prediction models, were also significant in this analysis. In the non-GHD group, (IGF-I – IGFBP-3) SDS was a positive predictor of growth but IGFBP-3 SDS was now a negative predictor, indicating that *readily available* IGF-I is important to achieve good growth.

Table 3. Range of diagnoses in the cross-sectional survey of IGF-I and IGFBP-3 monitoring while on GH treatment (Das et al. 2003).

GHD	Total **134**
Isolated	61
Hypopituitarism	41
Suprasellar tumours	19
Post-radiation	13
Turner syndrome	Total **54**
Non-GHD	Total **27**
Idiopathic short stature	7
Intra-uterine growth retardation	8
Syndromes with growth failure	6
Skeletal dysplasia	4
Chronic renal failure	2

We also examined those parameters influencing growth performance in tertiles of time on GH treatment in those with GHD (Table 6). BMI SDS was a significant determinant in the first tertile, whereas the IGF parameters became significant when the child had been on GH for a longer period.

Conclusions

The measurement of serum IGF-I and IGFBP-3 forms an important part of the assessment of the short, slowly growing child. However, such measurements are one component of a multi-faceted process, which includes history, examination, biochemical and radiological evaluation, leading to the diagnosis of GHD. A percentage of short, non-GHD children have IGF-I values comparable to those with GHD, and the present level of diagnostic performance is unlikely to be exceeded by any further refinements to the assay or cut-off values. However, assessment of IGF-I prohormone and E-peptide levels may provide further insight into the pericellular actions of IGF-I and may refine evaluation of GH-dependent IGF-I.

With regard to monitoring, high IGF-I and IGFBP-3 are not common in GHD but are more frequently seen in TS and SGA children treated with high doses of GH. Nevertheless, a significant number of IGF tests will indicate the need to consider dose adjustment. The determinants of growth in terms of IGF parameters are disease specific – (IGF-I – IGFBP-3) SDS is a positive determinant in GHD

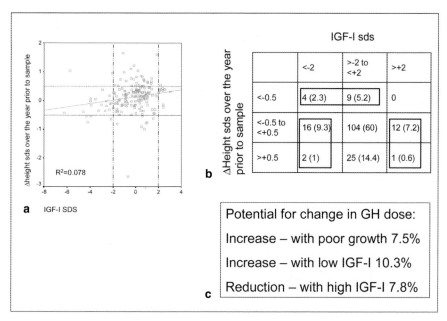

Fig. 3. A. The relationship between IGF-I SDS and change in height SDS over the year preceding the IGF-I measurement in GHD, TS and non-GHD children on GH treatment. B. The number and percentage of tests with high, low or normal IGF-I SDS values compared to growth performance. C. The percentage of tests where a change in management might be considered.

Table 4. Univariate regression analysis to define those parameters related to change in height SDS over the year up to IGF-I/IGFBP-3 measurement in GHD, TS and non-GHD children (Das et al. 2003).

Parameter	Variable	GHD	Turner Syndrome	Non-GHD
ΔHt sds over the year to sample	IGF-I SDS	R=+0.28, p=0.003	NS	NS
	IGFBP-3 SDS	NS	NS	NS
	(IGF-I - IGFBP-3 SDS	R=+0.23, p=0.03	NS	NS
	GH dose (mg/m²/d)	NS	R=-0.43, p=0.006	NS
ΔHt sds during GH treatment	GH dose (mg/m²/d)	R=+0.25, p=0.02	NS	NS

Table 5. Backward logistic regression to define those parameters related to change in height SDS over the year up to IGF-I/IGFBP-3 measurement in GHD, TS and non-GHD children (Das et al. 2003).

GHD group Variable	Standard Coefficient	P value
Midparental Ht sds – Ht sds at sample	0.214	0.094
BMI sds	0.292	0.016
IGFBP-3 sds	0.257	0.041
GH dose ($mg/m^2/d$)	0.225	0.063
[IGF-I – IGFBP-3] sds	0.337	0.007

Model r=0.45, P=0.012

Turner group Variable	Standard Coefficient	P value
Duration of GH treatment	-0.667	<0.001

Model r=0.67, P<0.001

Non-GHD group Variable	Standard Coefficient	P value
Age	-0.733	0.05
Birth weight	1.59	0.008
Midparental Ht sds	1.37	0.006
IGFBP-3 sds	-0.693	0.046
[IGF-I – IGFBP-3] sds	1.53	0.02
Midparental Ht sds – Ht sds at start of GH	-1.08	0.018

Model r=0.9, P<0.001

and non-GHD conditions, but IGFBP-3 SDS is positive in GHD and negative in non-GHD.

Both IGF-I and IGFBP-3 provide important information for monitoring, and thus both should be measured. Protocols for dose adjustments have not been defined and neither has the intensity of testing. The relationship between IGF-I and IGFBP-3 is complex and disease-specific and needs further study to better inform management of GH-treated patients.

Table 6. Backward logistic regression to define those parameters related to change in height SDS over the year up to IGF-I/IGFBP-3 measurement in GHD children in tertiles of time on GH treatment: <2.5 years, 2.5-5.7 years, >5.7 years. (MPH = mid-parental height sds).

Variable	GH <2.5 years		GH 2.5-5.7 years			GH >5.7 years		
	Standard Coefficient	P value	Variable	Standard Coefficient	P value	Variable	Standard Coefficient	P value
BMI sds	0.564	0.006	IGFBP-3 sds	0.354	0.085	IGFBP-3 sds	0.536	0.001
			MPH – Start Height sds	0.58	0.008	[IGF-I – IGFBP-3] sds	0.45	0.006
	r=0.56, p=0.006		r=0.62, p=0.016			r=0.67, p=0.001		

References

Aguiar-Oliveira MH, Gill MS, de A Barretto ES, Alcantara MR, Miraki-Moud F, Menezes CA, Souza AH, Martinelli CE, Pereira FA, Salvatori R, Levine MA, Shalet SM, Camacho-Hubner C, Clayton PE (1999) Effect of severe growth hormone (GH) deficiency due to a mutation in the GH-releasing hormone receptor on insulin-like growth factors (IGFs), IGF-binding proteins, and ternary complex formation throughout life. J Clin Endocrinol Metab 84:4118-126.

Boquete HR, Sobrado PG, Fideleff HL, Sequera AM, Giaccio AV, Suarez MG, Ruibal GF, Miras M (2003) Evaluation of diagnostic accuracy of insulin-like growth factor (IGF)-I and IGF-binding protein-3 in growth hormone-deficient children and adults using ROC plot analysis. J Clin Endocrinol Metab 88:4702-4708.

Chan JM, Stampfer MJ, Giovannucci E, Gann PH, Ma J, Wilkinson P, Hennekens CH, Pollak M (1998) Plasma insulin-like growth factor-l and prostate cancer risk: a prospective study. Science 279:563-566.

Clayton PE (1999) The role of insulin-like growth factors and other biochemical parameters in the diagnosis of growth hormone deficiency. In: Ranke MB, Wilton P (eds) Growth hormone therapy - ten years of KIGS 1987-1997. J.A. Barth Verlag, Heidelberg, pp 53-64.

Cohen P, Bright GM, Rogol AD, Kappelgaard AM, Rosenfeld RG, American Norditropin Clinical Trials Group (2002) Effects of dose and gender on the growth and growth factor response to GH in GH-deficient children: implications for efficacy and safety. J Clin Endocrinol Metab 87:90-98.

Das U, Whatmore AJ, Khosravi J, Wales JK, Butler G, Kibirige MS, Diamandi A, Jones J, Patel L, Hall CM, Price DA, Clayton PE (2003) IGF-I and IGF-binding protein-3 measurements on filter paper blood spots in children and adolescents on GH treatment: use in monitoring and as markers of growth performance. Eur J Endocrinol 149:179-185.

Drake WM, Coyte D, Camacho-Hubner C, Jivanji NM, Kaltsas G, Wood DF, Trainer PJ, Grossman AB, Besser GM, Monson JP (1998) Optimizing growth hormone replacement therapy by dose titration in hypopituitary adults. J Clin Endocrinol Metab 83:3913-3919.

Hankinson SE, Willett WC, Colditz GA, Hunter DJ, Michaud DS, Deroo B, Rosner B, Speizer FE, Pollak M (1998) Circulating concentrations of insulin-like growth factor-l and risk of breast cancer. Lancet 351:1393-1396.

Juul A, Bernasconi S, Clayton PE, Kiess W, DeMuinck-Keizer Schrama S, Drugs and Therapeutics Committee of the European Society for Paediatric Endocrinology (ESPE) (2002) European audit of current practice in diagnosis and treatment of childhood growth hormone deficiency. Horm Res 58:233-241.

Kuo YH, Chen TT (2002) Novel activities of pro-IGF-I E peptides: regulation of morphological differentiation and anchorage-independent growth in human neuroblastoma cells. Exp Cell Res 280:75-89.

Lanes R, Jakubowicz S (2002) Is insulin-like growth factor-1 monitoring useful in assessing the response to growth hormone of growth hormone-deficient children? J Pediatr 141:606-610.

Lowe WL Jr, Lasky SR, LeRoith D, Roberts CT Jr. (1988) Distribution and regulation of rat insulin-like growth factor I messenger ribonucleic acids encoding alternative carboxyterminal E-peptides: evidence for differential processing and regulation in liver. Mol Endocrinol 2:528-535.

Musaro A, Giacinti C, Borsellino G, Dobrowolny G, Pelosi L, Cairns L, Ottolenghi S, Cossu G, Bernardi G, Battistini L, Molinaro M, Rosenthal N (2004) Stem cell-mediated muscle regeneration is enhanced by local isoform of insulin-like growth factor 1. Proc Natl Acad Sci USA 101:1206-1210.

O'Sullivan DC, Szestak TA, Pell JM (2002) Regulation of IGF-I mRNA by GH: putative functions for class 1 and 2 message. Am J Physiol Endocrinol Metab 283:E251-258.

Petridou E, Dessypros N, Spanos E, Mantzoros C, Skalkidou A, Kalmanti M, Koliouskas D, Kosmidis H, Panagiotou JP, Piperopoulou F, Tzortzatou F, Trichopoulos D (1999) Insulin-like growth factor-I and binding protein-3 in relation to childhood leukemia. Int J Cancer 80:494-496.

Ranke MB, Schweizer R, Elmlinger MW, Weber K, Binder G, Schwarze CP, Wollmann HA (2001) Relevance of IGF-I, IGFBP-3, and IGFBP-2 measurements during GH treatment of GH-deficient and non-GH-deficient children and adolescents. Horm Res 55:115-124.

Report of a Workshop (2000) Consensus guidelines for the diagnosis and treatment of GH deficiency in childhood and adolescence: Summary statement of the GH Research Society on child and adolescent GH deficiency. J Clin Endocrinol Metab 85: 3990-3993.

Report of a Workshop (2001) Critical evaluation of the safety of recombinant human growth hormone administration: Statement from the Growth Hormone Research Society. J Clin Endocrinol Metab 86:1868-1870.

Rosenfeld RG (1997) Is growth hormone deficiency a viable diagnosis? J Clin Endocrinol Metab. 82:349-351.

Swerdlow AJ, Higgins CD, Adlard P, Preece MA (2002) Risk of cancer in patients treated with human pituitary growth hormone in the UK, 1959-85: a cohort study. Lancet 360:273-277.

Tillmann V, Patel L, Gill MS, Whatmore AJ, Price DA, Kibirige MS, Wales JK, Clayton PE (2000) Monitoring serum insulin-like growth factor-I (IGF-I), IGF binding protein-3 (IGFBP-3), IGF-I/IGFBP-3 molar ratio and leptin during growth hormone treatment for disordered growth. Clin Endocrinol (Oxf) 53:329-336.

van Pareren YK, de Muinck Keizer-Schrama SM, Stijnen T, Sas TC, Jansen M, Otten BJ, Hoorweg-Nijman JJ, Vulsma T, Stokvis-Brantsma WH, Rouwe CW, Reeser HM, Gerver WJ, Gosen JJ, Rongen-Westerlaken C, Drop SL (2003a) Final height in girls with Turner syndrome after long-term growth hormone treatment in three dosages and low dose estrogens. J Clin Endocrinol Metab 88:1119-1125.

Van Pareren Y, Mulder P, Houdijk M, Jansen M, Reeser M, Hokken-Koelega A (2003b) Adult height after long-term, continuous growth hormone (GH) treatment in short children born small for gestational age: results of a randomized, double-blind, dose-response GH trial. J Clin Endocrinol Metab 88:3584-3590.

Wyatt D (2004) Lessons from the national cooperative growth study. Eur J Endocrinol 151 Suppl 1:S55-59.

Yu H, Spitz MR, Mistry J, Gu J, Hong WK, Wu X (1999) Plasma levels of insulin-like growth factor-I and lung cancer risk: a case-control analysis. J Natl Cancer Inst 91:151-156.

IGF Resistance:
The Role of the Type 1 IGF Receptor

Steven D. Chernausek[1], M. Jennifer Abuzzahab[2], Wieland Kiess[3], Doreen Osgood[4], Anke Schneider[3] and Robert J. Smith[1]

Summary

The growth hormone (GH)-insulin-like growth factor (IGF)-I axis is the dominant regulator of somatic growth in vertebrates. Growth deficits of varying severity follow experimental deletion of individual IGF axis components, with all but about 15% of the growth of an adult mouse explicable by the combined contributions of growth hormone and IGF-I stimulation. The type 1 IGF receptor mediates the growth-promoting actions of the IGFs and, when absent in mice, leads to severe fetal growth retardation and perinatal lethality. Thus, the IGF receptor is a critical element along the GH/IGF stimulatory pathway. Specific defects of IGF receptor function have not been unequivocally demonstrated to cause disease in humans until recently. We have described two patients with mutations in the Type 1 IGF receptor gene associated with fetal and postnatal growth retardation. One child was a compound heterozygote for missense mutations in the ligand binding domain. These mutations lowered the affinity of the IGF receptor for IGF-I and attenuated receptor signaling. The other child had a stop codon in exon 2 and showed a reduction in cell surface IGF receptor abundance. Though specific mutations within the receptor gene will be uncommon causes of fetal and prenatal growth deficits, deficits in IGF-I signaling, either at the receptor or post-receptor level, will likely be implicated in a much wider variety of growth disorders. The identification of additional patients with partial defects in IGF receptor function and their careful phenotyping will lead to a better understanding of the breadth and depth of IGF-I action in humans.

[1] University of Cincinnati, School of Medicine, Cincinnati Children's Hospital Medical Center, Cincinnati, Ohio, USA
[2] Children's Hospitals and Clinics, St. Paul, Minnesota, USA
[3] Hospital for Children and Adolescents, Leipzig, Germany
[4] Hallet Center for Diabetes and Endocrinology, Brown Medical School, Providence, Rhode Island, USA

Carel et al.
Deciphering Growth
© Springer-Verlag Berlin Heidelberg

Introduction

Syndromes of hormone resistance have been recognized for decades. In several instances, the clinical phenotypes are well-characterized and the biochemical origins understood. The syndromes of insensitivity to growth hormone and to androgens caused by mutations in hormone receptor genes are apt examples. Patients with insulin resistance due to insulin receptor mutations show a range of clinical manifestations that correlate with the degree of receptor dysfunction (Accili 1995; Taylor et al. 1994). Though such patients are uncommon, insights from studies of them have had broad implications because insulin resistance, at some level, is a characteristic feature of type II diabetes mellitus. Resistance to the insulin-like growth factors (IGFs) has received less attention owing to the complexity of the IGF system, the lack of documented genetic syndromes of IGF resistance, and intrinsic difficulties in studying certain aspects of the IGF system in humans. Because the network of components involved in growth control is vast and the effects of IGF are wide ranging, all defects of growth not due to IGF deficiency might be considered forms of IGF resistance. A somewhat restricted definition is therefore helpful to maintain focus. In this work, we will consider IGF resistance as an impairment of a cellular or organ response to IGF that is 1) restricted to actions directly attributable to IGF-I or IGF-II and 2) occurs in the presence of adequate IGF production. Such a definition includes genetic mutations of the type 1 IGF receptor (IGF 1R), post-receptor defects in the more proximal components involved in IGF 1R signaling, and reductions in IGF action due to an excess of inhibitory IGF binding proteins. It would extend to situations where a mutant IGF acts to block IGF 1R activation. Excluded are defects primarily attributable to reduced growth hormone action and intrinsic, core defects in cellular growth regulation, such as abnormalities of proteins that regulate cell cycle length. This review will focus on genetic causes of IGF resistance, specifically those that result from defects in the IGF receptor. Nevertheless, many of the conclusions are likely to be relevant to acquired states of IGF resistance that undoubtedly develop in a variety of pathologic circumstances.

Roles of IGF as modulators of growth and metabolism

The IGFs are the dominant regulators of somatic growth and no organ is free from their influence. The IGFs help determine the size of tissues, stimulating cell growth and replication. However, the roles of IGF extend beyond control of growth to the regulation of tissue-specific functions. Functions of the IGFs can be conceptually separated into three categories. The first is the general regulation of growth that occurs because IGFs increase cell size by promoting protein synthesis and increase cell number by the simultaneous stimulation of cell replication and attenuation of apoptosis (Popken et al. 2004). The second category involves direct action of IGFs, separate from those involved in growth. Examples include the simulation of glucose uptake (an "insulin-like" effect) and enhancement of bone mineralization due to direct actions on osteoblasts (Clemens and Chernausek 2004). The third category involves actions where IGF tone determines the magnitude of the effects of another regulatory molecule. For example,

Table 1. Tissues in which IGFs are known to act and postulated actions. The data are derived largely from in vitro studies and experiments using animal models.

IGF actions	
• Skeletal growth	• Kidney growth
• Bone remodeling and mineralization	• Gut growth
• Brain growth and myelination	• Cardiovascular growth and function
• Breast development and lactation	• Promote neoplastic growth
• Insulin secretion and action	• Steroidogenesis
	• Placental growth and function

in the absence of IGF-I, follicle-stimulating hormone (FSH) only weakly stimulates granulosa cell steroidogenesis. When IGF-I is present, however, the ability for FSH effect is markedly enhanced (Adashi 1998; LaVoie et al. 1999).

The breadth of tissues responsive to IGFs and the variety of mechanisms implicated suggest important involvement of the IGFs in many physiologic processes (see Table 1). In the context of such widespread action, it is difficult to predict the phenotype in circumstances of IGF 1R signal attenuation. Indeed, different tissues may have distinct sensitivities to reduction in IGF signaling or variable abilities to compensate. Thus, the identification and study of IGF-resistant humans, as well as the creation of experimental models wherein IGF resistance is produced, provide important insights into the physiologic roles of IGF and begin to define the phenotype of IGF resistance.

Phenotypes of experimental IGF I resistance

Expectations for clinical features of IGF resistance in humans come mainly from studies of mice in which specific components of the IGF axis have been experimentally deleted or modified (Efstratiadis 1998; Lupu et al. 2001). In the mouse, two IGFs (I and II) are secreted during fetal life. IGF II plays the dominant role in early fetal growth but, shortly following birth, its generalized expression declines and it disappears from the bloodstream. It is replaced by IGF I, which assumes an increasingly important role in growth control beginning late in fetal life and continuing throughout the growing period of the mouse. Even though two distinct IGFs are involved in growth regulation, they both act through the type I IGF receptor (IGF 1R), a disulfide-linked heterotetramer homologous with the insulin receptor (Dupont and LeRoith 2001; LeRoith et al. 1995). The effects of various mutations in the IGF signal transduction pathway on murine growth are illustrated in Figure 1. The absence of IGF binding proteins (IGFBPs) has little effect on overall somatic growth (Pintar et al. 1997). Likewise, the removal of a single element in the intracellular signal transduction pathway typically has only a modest growth-retarding effect (Fantin et al. 2000; Kadowaki et al. 1996;

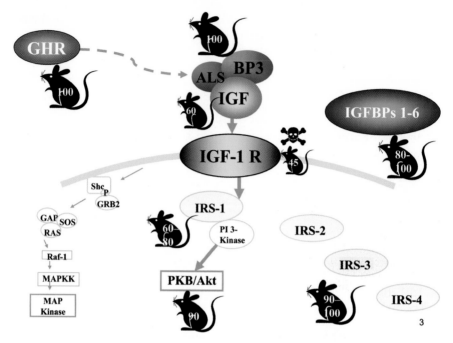

Fig. 1. Fetal growth effects of lesions in the GH/IGF axis. The approximate size of mice born with null mutations in specific genes is indicated as a percentage of wild-type birth weight. Note that defects in the type I IGF-I receptor and its respective ligand result in the smallest mice, with perinatal lethality in the IGF 1 R null mutant. Data are from a variety of studies cited in the text. GHR, growth hormone receptor; ALS, acid labile subunit; BP3, insulin-like growth factor binding protein (IGFBP) 3; IRS, insulin receptor substrate.

Liu et al. 1999; Peng et al. 2003; Tamemoto et al. 1994). Thus, it seems that the complex network that regulates access of the IGFs to the IGF 1R, as well as the interactions between intracellular signaling pathways, allows for physiologic adjustments that can largely compensate for the deficiency, at least as far as overall growth is concerned. In contrast, IGF 1R null mutant mice are extremely small at birth, demonstrating that the receptor is critically positioned as the gateway through which the extracellular signals must pass to stimulate a cascade of intracellular events that ultimately result in cell growth and division. It follows, therefore, that both genetic and acquired disorders that impinge on the IGF 1R or the initial post-receptor signaling events will have profound effects on growth.

For the most part, the murine system is a good model for forming and testing hypotheses concerning IGF in man. All the major components of the GH/IGF axis are present and highly homologous. However, there are some distinctions between mice and men in terms of specific regulations of the different components. For example, levels of IGF-II are maintained in the circulation in man, but not in the mouse; the dynamics of GH release differ as well (Douyon and Schteingart 2002; Giustina and Veldhuis 1998). Thus, hypotheses concerning the roles of various elements of the GH/IGF axis that are derived from animal models should be tested by examining humans with equivalent genetic defects.

IGF I resistance in humans

Previous studies have suggested that IGF resistance explains the growth deficits in certain human conditions. The African Efe Pygmy is very small as an adult, despite relatively normal circulating levels of IGF-I. Studies by Geffner et al. (1995; Hattori et al. 1996) have implied the Pygmies' aberrant growth is due to impairment in IGF 1R function; however, the precise molecular mechanism remains to be elucidated. Growth abnormalities of individuals with gross defects of chromosome 15 also potentially involve the IGF 1R. In humans, the type 1 IGF receptor gene is located on the distal end of the long arm of chromosome 15. Several patients have been reported in whom the portion of the chromosome bearing the IGF 1R gene has been lost, due to terminus deletion, translocation, or ring formation.(Okubo et al. 2003; Peoples et al. 1995; Roback et al. 1991). These patients typically experience intrauterine growth retardation (IUGR) and are short. Some have increased serum IGF-I concentrations, implying resistance to the IGFs and suggesting the growth phenotype is due to a reduction in IGF 1R gene dosage. Dysmorphic features and severe mental retardation are found as well. Because many other genes are disrupted in addition to the IGF 1R, it is difficult to know which aspects of the phenotype are attributable to an IGF-I signaling deficit and which reflect separate anomalies.

To better understand the consequences of IGF resistance in humans, we began to search for patients with abnormalities of IGF 1R function. Because prenatal growth retardation was such a prominent feature of mice with IGF 1R deficiency, we initially examined a group of patients who had short stature and IUGR,using single strand conformation polymorphism analysis to screen for genetic abnormalities of the IGF 1R. In a parallel study, colleagues examined the registry of short patients, sequencing of the IGF 1R gene in individuals with relatively high circulating IGF-I concentrations. These efforts identified the first two patients with IGF resistance due to IGF receptor mutations (Abuzzahab et al. 2003). In one case the patient was a compound heterozygote for missense mutations that altered the amino acid sequence within the ligand binding domain (arg108gln/lys115asp). This patient had reduced IGF-I binding and her receptors were less sensitive to stimulation by IGF-I. She had high circulating concentrations of IGF-I, IGFBP-3 and the acid labile subunit (ALS), accompanied by elevated measures of growth hormone secretion, presumably resulting from a lack of IGF-I feedback at the somatotroph. The other case had a nonsense mutation in exon 2 that decreased the abundance of cell surface IGF 1R. He also had IUGR and short stature.

It is of interest to compare these recently described patients with others with different genetic defects of GH and IGF action. (Table 2) The dominant role of IGF in the control of fetal growth is clear, with patients with primary deficiency of IGF-I or IGF 1R deficiency displaying severe IUGR (Abuzzahab et al. 2003; Woods et al. 1996). In contrast, children with GH insensitivity have near normal prenatal growth (Rosenbloom et al. 1997). The potential role of IGF-I in carbo-

Table 2. Comparison between patients with diverse genetic defects.

	GH R D[a]	GH Post-R (Stat 5-b)	IGF-I D	IGF 1R D
GH secretion	Increased	Increased	Increased	Varied?
Serum IGF-I	Very low	Very low	Absent	Nl to Inc
Serum IGF-II	Low	n/a	Normal	Normal
IGFBP-3	Very low	Very low	Normal	Nl to Inc
Prenatal growth	Near normal	Near normal	IUGR	IUGR
Postnatal growth	Very slow	Very slow	Very slow	Very slow
Skeletal material	Very delayed	n/a	Modest delay	Modest delay
CNS	Near normal	n/a	Retarded	Variably abnormal
Hearing	Normal?	n/a	Sensorineural deafness	Normal?
Glycemic status	Hypoglycemia	n/a	CHO Intol	CHO Intol
Dysmorphic features	Frontal bossing Mid-face hypoplasia	Frontal bossing Mid-face hypoplasia	Yes	Variable
Immunologic status	Clinically normal	Impaired, with frequent infections	Clinically normal	Clinically normal

[a] GH R D, GH insensitivity due to GH receptor deficiency (Laron syndrome); GH R post-R, Stat 5 b deficiency (single case report; Kofoed et al. 2003); IGF-I D, deficiency of IGF-I gene (single case; Woods et al. 1996); IGF 1R D, genetic mutations in type I IGF receptor; Abuzzahab et al. 2003; Kawashima et al. 2003)

hydrate homeostasis is also intriguing, suggested by abnormalities of CHO metabolism described in patients with defects in the IGF pathway (Sundararajan et al. 2004; Woods et al. 2000). Indeed, studies in mice indicate that IGF 1R signaling is essential for normal – cell function (though apparently not – cell growth; Kido et al. 2002; Kulkarni et al. 2002; Xuan et al. 2002).

Though the depth and breadth of the phenotype of IGF resistance will require more study and necessitate identification of additional patients, a picture of IGF resistance is beginning to emerge. The two patients reported had relatively increased levels of IGF-I in circulation, though not always out of the normal range. Growth failure that begins during the prenatal period may be characteristic, as might be adverse effects on CNS function leading to mental retardation, microcephaly, or psychiatric disturbance. The absence of major organ malformations and severe dysmorphism is in keeping with the role of IGF-1R as a primary regulator of growth rather than differentiation. However, the two cases we reported, as well as a preliminary report of a growth-retarded child heterozygous for a defect in the cleavage region between the α and β subunits of the receptor (Kawashima et al. 2003), all have residual IGF-1R function. An IGF 1R null lesion may not be compatible with life.

Attenuated forms of IGF resistance

From these case studies it is evident that defects that reduce but do not eliminate IGF 1R function can exert profound effects. (The mature height for the patient with the arg108gln/lys115asp mutations was 134 cm, or -4.8 SD for normal women.) Furthermore, there is clear evidence for heterozygote effect in the proband and family members carrying a stop codon in exon 2 in one allele. [This is perhaps another distinction between species; mice hemizygous null for the IGF 1R generally show normal growth (Liu et al. 1993).] Thus, it is worthwhile to reflect on the potential biological effects of more modest reductions in IGF-1R signaling. The heterotetrameric structure of the IGF 1R may predispose it to dominant negative effects from expressed, mutant receptors. Consider the IGF 1R of the father of the heterozygous arg108gln/lys115asp patient. He is healthy and has normal circulating IGF-I levels, but his adult height is -2.8 SD. If the IGF 1R protein is produced in equivalent amounts from each allele, only 25% of cell-surface receptors will be normal (Fig. 2). If it takes a completely normal IGF 1R to function, this effect could be significant. If such mechanisms are operative, heterozygous mutations in the IGF 1R gene could have meaningful effects in a much larger group of humans than expected.

Conclusions and Speculation

- IGF resistance due to mutations of the IGF 1R gene is characterized by pre- and postnatal growth failure.
- The growth abnormalities found in chromosome 15 aneuploidy are at least partly due to the effects of IGF 1R gene dosage.

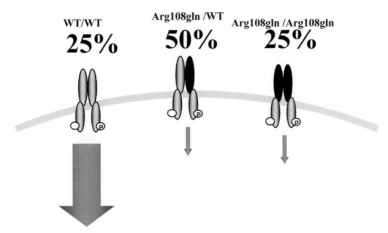

Fig. 2. Hypothetical effect of heterozygous mutation (arg108gln, indicated by black color) in IGF 1R gene. Because the mature IGF 1R is a heterotetramer composed of subunits potentially derived from either allele, only 25% of formed, cell surface receptors will be free of the mutation as illustrated. If only receptors with two wild-type α subunits function normally, a substantial reduction in net IGF 1R signal would be expected.

- Graded reductions in IGF 1R signaling produce a range of clinical manifestations that will be determined, in part, by the extent to which IGF 1R function is affected.
- Hypomorphic lesions in the IGF-1R and the proximal elements of IGF-1R signal transduction are likely to be more common than anticipated and cause moderate short stature, abnormalities of carbohydrate metabolism, and mental status variation.

References

Abuzzahab MJ, Schneider A, Goddard A, Grigorescu F, Lautier C, Keller E, Kiess W, Klammt J, Kratzsch J, Osgood D, Pfaffle R, Raile K, Seidel B, Smith RJ, Chernausek SD (2003) IGF-I receptor mutations resulting in intrauterine and postnatal growth retardation. New Engl J Med 349:2211-2222

Accili D (1995) Molecular defects of the insulin receptor gene. Diabetes/Metab Rev 11:47-62

Adashi EY (1998) The IGF family and folliculogenesis. J Reprod Immunol 39:13-19

Clemens TL, Chernausek SD (2004) Genetic strategies for elucidating insulin-like growth factor action in bone. Growth Horm IGF Res 14:195-199

Douyon L, Schteingart DE (2002) Effect of obesity and starvation on thyroid hormone, growth hormone, and cortisol secretion. Endocrinol Metab Clin North Am 31:173-189

Dupont J, LeRoith D (2001) Insulin and insulin-like growth factor I receptors: similarities and differences in signal transduction. Horm Res 55 Suppl 2:22-26

Efstratiadis A (1998) Genetics of mouse growth. Int J Dev Biol 42:955-976.

Fantin VR, Wang Q, Lienhard GE, Keller SR (2000) Mice lacking insulin receptor substrate 4 exhibit mild defects in growth, reproduction, and glucose homeostasis. Am J Physiol Endocrinol Metab 278:E127-133

Geffner ME, Bersch N, Bailey RC, Golde DW (1995) Insulin-like growth factor I resistance in immortalized T cell lines from African Efe Pygmies. J Clin Endocrinol Metab 80:3732-3738

Giustina A, Veldhuis JD (1998) Pathophysiology of the neuroregulation of growth hormone secretion in experimental animals and the human. Endocr Rev 19:717-797.

Hattori Y, Vera JC, Rivas CI, Bersch N, Bailey RC, Geffner ME, Golde DW (1996) Decreased insulin-like growth factor I receptor expression and function in immortalized African Pygmy T cells. J Clin Endocrinol Metab 81:2257-2263

Kadowaki T, Tamemoto H, Tobe K, Terauchi Y, Ueki K, Kaburagi Y, Yamauchi T, Satoh S, Sekihara H, Aizawa S, Yazaki Y (1996) Insulin resistance and growth retardation in mice lacking insulin receptor substrate-1 and identification of insulin receptor substrate-2. Diabet Med 13:S103-108

Kawashima Y, Kinoshita T, Hanaki K, Nagaishi J, Nagata I, Y N, Ohtsuka Y, Hisatome I, Ninomiya H, Nanba E, Kanzaki S. (2003) The missense mutation at the cleavage site of the insulin-like growth factor I receptor (arg709gln) in a family with short stature born intrauterine growth retardation. Proc Endocr Soc 85th Annual Meeting : 393

Kido Y, Nakae J, Hribal ML, Xuan S, Efstratiadis A, Accili D (2002) Effects of mutations in the insulin-like growth factor signaling system on embryonic pancreas development and beta-cell compensation to insulin resistance. J Biol Chem 277:36740-36747

Kofoed EM, Hwa V, Little B, Woods KA, Buckway CK, Tsubaki J, Pratt KL, Bezrodnik L, Jasper H, Tepper A, Heinrich JJ, Rosenfeld RG (2003) Growth hormone insensitivity associated with a STAT5b mutation. New Engl J Med 349:1139-1147

Kulkarni RN, Holzenberger M, Shih DQ, Ozcan U, Stoffel M, Magnuson MA, Kahn CR (2002) beta-cell-specific deletion of the Igf1 receptor leads to hyperinsulinemia and glucose intolerance but does not alter beta-cell mass. Nature Genet 31:111-115

LaVoie HA, Garmey JC, Veldhuis JD (1999) Mechanisms of insulin-like growth factor I augmentation of follicle-stimulating hormone-induced porcine steroidogenic acute regulatory protein gene promoter activity in granulosa cells. Endocrinology 140:146-153

LeRoith D, Werner H, Beitner-Johnson D, Roberts CT, Jr (1995) Molecular and cellular aspects of the insulin-like growth factor I receptor. Endocr Rev 16:143-163

Liu J-P, Baker J, Perkins AS, Robertson EJ, Efstratiadis A (1993) Mice carrying null mutations of the genes encoding insulin-like growth factor I (Igf-1) and the type 1 IGF receptor (Igf1r). Cell 75:59-72

Liu SC, Wang Q, Lienhard GE, Keller SR (1999) Insulin receptor substrate 3 is not essential for growth or glucose homeostasis. J Biol Chem 274:18093-18099

Lupu F, Terwilliger JD, Lee K, Segre GV, Efstratiadis A (2001) Roles of growth hormone and insulin-like growth factor 1 in mouse postnatal growth. Dev Biol 229:141-162.

Okubo Y, Siddle K, Firth H, O'Rahilly S, Wilson LC, Willatt L, Fukushima T, Takahashi S, Petry CJ, Saukkonen T, Stanhope R, Dunger DB (2003) Cell proliferation activities on skin fibroblasts from a short child with absence of one copy of the type 1 insulin-like growth factor receptor (IGF1R) gene and a tall child with three copies of the IGF1R gene. J Clin Endocrinol Metab 88:5981-5988

Peng XD, Xu PZ, Chen ML, Hahn-Windgassen A, Skeen J, Jacobs J, Sundararajan D, Chen WS, Crawford SE, Coleman KG, Hay N (2003) Dwarfism, impaired skin development, skeletal muscle atrophy, delayed bone development, and impeded adipogenesis in mice lacking Akt1 and Akt2. Genes Dev 17:1352-1365

Peoples R, Milatovich A, Francke U (1995) Hemizygosity at the insulin-like growth factor I receptor (IGF1R) locus and growth failure in the ring chromosome 15 syndrome. Cytogenet Cell Genet 70:228-234

Pintar JE, Schuller A, Bradshaw S, Cerro J, Wood T (1997) Genetic analysis of IGFBP function. Proc Endocr Soc 79th Annual Meeting:59

Popken GJ, Hodge RD, Ye P, Zhang J, Ng W, O'Kusky JR, D'Ercole AJ (2004) In vivo effects of insulin-like growth factor-I (IGF-I) on prenatal and early postnatal development of the central nervous system. Eur J Neurosci 19:2056-2068

Roback EW, Barakat AJ, Dev VG, Mbikay M, Chrétien M, Butler MG (1991) An infant with deletion of the distal long arm of chromosome 15 (q26.1-->qter) and loss of insulin-like growth factor 1 receptor gene. Am J Med Genet 38:74-79

Rosenbloom AL, Rosenfeld RG, Guevara-Aguirre J. (1997) Growth hormone insensitivity. Ped Clin N Am 44:423-442

Sundararajan S, Banach W, Chernausek SD (2004) Defective Glucose homeostasis in a child with a genetically-acquired function defect in the IGF-I receptor. Proc Ped Soc Annual Meeting : 881

Tamemoto H, Kadowaki T, Tobe K, Yagi T, Sakura H, Hayakawa T, Terauchi Y, Ueki K, Kaburagi Y, Satoh S, Sekihara H, Yoshioka S, Horikoshi H, Furuta Y, Ikawa Y, Kasuga M, Yazaki Y, Aizawa S (1994) Insulin resistance and growth retardation in mice lacking insulin receptor substrate-1. Nature 372:182-186

Taylor SI, Accili D, Haft CR, Hone J, Imai Y, Levy-Toledano R, Quon MJ, Suzuki Y, Wertheimer E (1994) Mechanisms of hormone resistance: lessons from insulin-resistant patients. Acta Paediatr.Suppl 399:95-104

Woods KA, Camacho-Hubner C, Savage MO, Clark AJL (1996) Intrauterine growth retardation and postnatal growth failure associated with deletion of the insulin-like growth factor I gene. New Engl J Med 335:1363-1367

Woods KA, Camacho-Hubner C, Bergman RN, Barter D, Clark AJ, Savage MO (2000) Effects of insulin-like growth factor I (IGF-I) therapy on body composition and insulin resistance in IGF-I gene deletion. J Clin Endocrinol Metab 85:1407-1411

Xuan S, Kitamura T, Nakae J, Politi K, Kido Y, Fisher PE, Morroni M, Cinti S, White MF, Herrera PL, Accili D, Efstratiadis A (2002) Defective insulin secretion in pancreatic beta cells lacking type 1 IGF receptor. J Clin Invest 110:1011-1019

The Importance
of the National Cooperative Growth Study (NCGS)

Raymond L. Hintz[1]

Introduction

Because all of the possible side effects of a drug cannot be anticipated based on preapproval studies involving only several dozen to several thousand patients, the U.S. Food and Drug Administration (FDA) maintains a system of post-marketing surveillance and risk assessment programs to identify adverse events that did not appear during the drug approval process. Companies are mandated to report at least yearly to the FDA, and more frequently if any safety concerns are detected. At the time of approval of recombinant human growth hormone in 1985, the FDA mandated Genentech to follow at least 300 patients for five years; because of this mandate, the National Cooperative Growth Study (NCGS) was started by Genentech. This original study has grown over the past two decades to include more than 50,000 treated patients and 171,000 patient years.

History

Genentech was founded in 1976, and one of its first projects was to develop a bio-recombinant synthetic human growth hormone (rhGH). GH was in chronically short supply because only human GH was effective in treating patients with GH deficiency and severe short stature. This species specificity mandated the use of human cadaver pituitary glands as the source for purifying GH for clinical use. Because of the shortage of donations of cadaver GH, it was estimated that less than half of the children with GH deficiency could be treated with GH. Imagine the situation if insulin action had been found to be species specific. Very few, if any, diabetic patients could have been treated with insulin until the development of synthetic insulin in the late twentieth century, instead of the therapeutic use of pork or beef insulin starting in 1922.

Clinical trials with Protropin (recombinant methionyl GH) began in 1981, initially in adult volunteers (Hintz et al. 1982), then proceeded quickly to studies in children with GH deficiency. The trials in children with GH deficiency were extremely successful (Kaplan et al. 1986), and Protropin was approved by the

---

[1] Department of Pediatrics, Stanford University Medical Center, Rm S-302 - 300 Pasteur Drive, Palo Alto, CA 94305, California, USA

Carel et al.
Deciphering Growth
© Springer-Verlag Berlin Heidelberg

Table 1. Demographics of patients in NCGS.

	1986 (%)	2004 (%)
Idiopathic GH deficiency	39	43
Organic GH deficiency	26	16
Idiopathic short stature	14	17
Turner Syndrome	7	10
Other, renal	14	14

FDA in October 1985 for use in children with "inadequate GH." This approval was timely since human pituitary GH use had been discontinued in the US and most of the world because of the occurrence of Creutzfeldt-Jakob disease in some young recipients of pituitary GH in the spring of 1985 (Hintz 1995). The FDA approval was contingent on doing a post-marketing study, and The National Cooperative Growth Study (NCGS) was set up simultaneously. The original FDA mandate was for a study of 300 patients for five years. By the fall of 1986. at the time of the first NCGS investigators' meeting, there were more than 1,100 patients enrolled in the study (Sherman 1987; Wyatt 2004).

The NCGS Today

There are now 11,000 patients being followed by more than 900 investigators in 381 active sites in North America. This total represents approximately 75% of all children being treated with Genentech GH. The NCGS study is still dynamic and growing, with 2,907 new patients and 30,029 patient visits in the past 12 months. More than 70 papers have been published by members of the NCGS team. These publications are listed chronologically in Appendix 1 in this paper. The NCGS patients are enrolled at the start of GH treatment and followed through the course of therapy, and post-treatment heights are measured until epiphyseal closure. Demographic and medical information is collected at the baseline and at each visit, every three to four months. This information includes height, weight, and pubertal status, bone age, injection site examinations, treatment regimen, concomitant medications, endocrine studies, and reportable adverse events. The most frequent diagnosis in the NCGS is still idiopathic GH deficiency, but many conditions are represented (Table 1).

The inclusion criteria for the NCGS are children of either sex who are treated with Nutropin AQ, Nutropin, or Protropin for the treatment of growth failure and who are compliant with visits. The exclusion criteria include:
- subjects treated with non-Genentech GH preparation
- subjects with closed epiphyses
- subjects with active neoplasia.

Intracranial lesions must be inactive and antitumor therapy must be completed for a period of 12 months prior to institution of GH therapy. The data gathered in the core protocol include:
- baseline demographics, as outlined on the Enrollment Form (Form 1) in a computerized database (GT Plus).
- information regarding growth history and history of any GH treatment (mandatory)
- baseline left hand/wrist X ray for bone age determination
- funduscopic examination
- medical history, laboratory measurements, and current clinical status, including height, weight, pubertal status, concomitant medications, treatment dose, and schedule.

The recommended schedule for study follow-up visits is every three to four months. The following data is recorded at each return visit in the computerized database:
- height, weight, and pubertal status (mandatory at first dose)
- bone age
- injection site examinations
- Nutropin AQ, Nutropin, or Protropin treatment regimen (mandatory)
- concomitant medications, endocrine studies
- subject compliance
- reportable adverse events.

NCGS Substudies

In addition to the NCGS Core Study, a total of 11 other substudies have been organized to answer specific questions. Seven of the substudies are now completed:
- Substudy 2: Serial GH sampling in the diagnosis of GH deficiency
- Substudy 3: Psychological testing
- Substudy 4: Daily dosing of GH
- Substudy 5: Urinary GH testing in the diagnosis of GH deficiency
- Substudy 6: GHBP, IGF-I, IGFBP-3
- Substudy 7: Adolescent bone age
- Substudy 8: Short stature.

Four of the NCGS substudies are still active:
- Substudy 9: Turner syndrome: includes girls with Turner syndrome who are treated with GH for growth failure. As of June 2004, there are 1,497 total patients with Turner syndrome enrolled in this study. Data collected provide information regarding growth history and history of GH treatment, and include:
  - baseline left hand & wrist X-Ray for bone age determination
  - funduscopic examination
  - medical history

- karyotype characteristics
- features of Turner syndrome
- physical findings
- family history of diabetes mellitus
– Substudy 10: Bone mineral density: includes adolescent boys and girls with GHD who have been enrolled and been followed on GH treatment in the NCGS Core protocol. It also includes adolescent girls with Turner syndrome who have been enrolled and followed during GH treatment in the NCGS core protocol. As of June 2004, this substudy included 55 total patients. The collected data include:
  - NCGS Core Study Discontinuation Form
  - original DEXA report (hologic or lunar)
  - bone age within six months of DEXA
– Substudy 11: Chronic renal insufficiency (CRI): includes children diagnosed with CRI or end-stage renal disease (ESRD) and treated with GH after January 2001. As of June 2004, this substudy had 139 patients enrolled. Collected data include:
  - screening/baseline evaluations
  - patient characteristics, medical and growth history
  - height of biological parents
  - etiology of renal disease and history of maintenance dialysis
  - chronic concomitant medications
  - physical examination to include height, weight, Tanner stage, BP, Serum creatinine
  - GH treatment plan
  Follow up evaluations in this substudy include:
  - height, weight, BP, serum creatinine
  - examination of injection sites, compliance to GH treatment plan
  - adverse events related or not related to GH therapy
  - concomitant medication change
– Substudy 12; Optimal GH dosing: includes adolescent GH-deflcient subjects, Tanner Stage 2 or greater, who are being treated or have been treated with GH to improve growth. There are 311 total patients as of August 2004. The collected data in addition to the Core protocol include IGF-I and IGFBP-3 serum levels at baseline and every three months.

NCGS Targeted Events

In addition to the problem-focused NCGS substudies, the NCGS database has been used to study targeted safety events that might be related to GH therapy. These include leukemia and other malignancies, diabetes mellitus, intracranial hyperglycemia and acute adrenal insufficiency.

Leukemia: Since a report from Japan in 1989 (Hara et al. 1989), there has been concern about the possible association of GH treatment and the development or recurrence of leukemia. The NCGS database has yielded reassuring evidence that this is not a risk in GH-treated patients without other risk factors such as radiation, chemotherapy or a syndrome with known predilection to develop leukemia (Allen et al. 1997; Taha et al. 2001) . A recent article from Japan reviewing all GH-treated patients in that country has also concluded that the risk in GH-treated patients of developing leukemia is not significantly different from the general population unless there are other risk factors (Nishi et al. 1999).

An analysis of the NCGS database also shows that the risk of developing *extracranial* nonleukemic malignancies is not increased in patients treated with GH (Blethen et al. 1996; Maneatis et al. 2000). This conclusion was recently reaffirmed by Sklar and coworkers (2002), who reviewed the data in the Childhood Cancer Survivor Study. The situation in GH-treated patients in terms of the risk of developing *diabetes mellitus* and *hyperglycemia* is less clear. Reduction of insulin sensitivity is a physiologic effect of GH; however, glucose homeostasis is maintained in the vast majority of patients treated with GH. Most of the available surveillance data do not demonstrate an increased incidence of diabetes, either type 1 or type 2, associated with GH treatment. There are, however, subgroups of patients (e.g., Turner's syndrome, Prader-Willi syndrome, and intrauterine growth retardation) who are inherently at risk of developing diabetes, and these should be carefully monitored (Statement from the Growth Hormone Research Society 2001). Analysis of the NCGS database suggests that there may be a slight increase in the standard morbidity rate (Maneatis et al. 2000). More studies will need to be done to settle this issue.

The association of *intracranial hypertension* with GH therapy was first discovered because of data reported by Genentech to the FDA (Malozowski et al. 1995). This association has since been validated worldwide in a variety of databases. This uncommon complication of GH therapy is reversible and in most cases GH therapy can be continued or resumed after a period off therapy.

The potential association of *acute adrenal insufficiency* with GH therapy was first noted in a long-term follow-up study of pituitary GH-treated patients in Canada (Taback and Dean 1996; Hintz 1996), where a surprisingly high proportion of the deaths in their study (9/37) was caused by the preventable endocrine complications of adrenal crisis and hypoglycemia. In 2002 Swerdlow and coworkers (2002) published a cohort study to investigated 1848 patients in the UK who were treated during childhood and early adulthood with human pituitary growth hormone during the period from 1959 to 1985. Although based on small numbers, they found a increased risk of colorectal cancer was of some concern and the authors concluded that further investigation in other cohorts was needed. Recently Mills and coworkers (2004) did a retrospective study of long-term mortality in the United States cohort of children treated with pituitary GH between 1963 and 1985. Of the 6,107 patients with 105,797 person-years of follow-up in this cohort, there were 433 deaths. In the general population, there would be 144 deaths expected for a relative risk of 3.8 (95% CI. 3.4 - 4.2). Of the 433 deaths, 106 were considered sudden and unexpected, and of these, 59/106 (56%) had clinical pictures suggesting acute adrenal insufficiency. They did not find an increased risk of colorectal cancer, unlike the UK study (Swerdlow et al. 2002) These findings reemphasize the importance of careful management of these patients into adult life and point out the importance of the NCGS in evaluating potential associated events.

Importance of the NCGS: Conclusions

Participation in the NCGS has been encouraged by frequent feedback of data to the investigators, by annual NCGS investigator meetings since 1986 and by the development of computerized data entry tools. The obvious strength of this study is the tremendous number of patients who participate, which generates data on a large proportion of the patients treated with a growth hormone. The analysis of NCGS data has provided reassuring evidence that GH is safe and effective in a wide variety of situations and has helped define the level of risk of certain rare complications. The NCGS also has the effect of providing a "snapshot" of pediatric endocrine practice in North America as it is happening in the real world.

The reservations on this or any other post-marketing study are that the data are the result of passive reporting, so that under-reporting and reporting/recall biases probably exist. The numbers are less certain than in a prospective study, and the information is frequently incomplete. In addition, there are no concurrent controls. These factors confound the assessment of association and causality. Nonetheless, the NCGS has contributed extremely valuable data on the safety and efficacy of biosynthetic GH products and has served as a focus of pediatric endocrine clinical investigation in North America.

References

Allen DB, Rundle AC, Graves DA, Blethen S (1997). Risk of leukemia in children treated with human growth hormone: review and reanalysis. J Pediatr 131:S32-36.

Blethen SL, Allen DB, Graves D, August G, Moshang T, Rosenfeld R (1996) Safety of recombinant deoxyribonucleic acid-derived growth hormone: The National Cooperative Growth Study experience. J Clin Endocrinol Metab 81:1704-1710.

Critical Evaluation of the Safety of Recombinant Human Growth Hormone Administration: Statement from the Growth Hormone Research Society.(2001) J Clin Endocrinol Metab 86:1868-1870.

Hara T, Komiyama A, Ono H, Akabane T (1989) Acute lymphoblastic leukemia in a patient with pituitary dwarfism under treatment with growth hormone. Acta Paediatr Jpn 31:73-77.

Hintz RL (1995) The prismatic case of creutzfeldt-jakob disease associated with pituitary growth hormone treatment. J Clin Endocrinol Metab 80:2298-2301.

Hintz RL (1996) Eternal vigilance--mortality in children with growth hormone deficiency. J Clin Endocrinol Metab 81:1691-1692, 1996.

Hintz RL, Rosenfeld RG, Wilson DM, Bennett A, Fenno J, McClellan B, Swift R (1982) Biosynthetic methionyl-human growth hormone is biologically active in adult man. Lancet I:1276-1279.

Kaplan SL, Underwood LE, August GP, Bell JJ, Blethen SL, Blizzard RM, Brown DR, Foley TP, Hintz RL, Hopwood NJ, Johansen A, Plotnik LP, Underwood NJ, Kirkland RT, Rosenfeld RG, Vanwyk JJ (1986) Clinical studies with recombinant-DNA-derived methionyl human growth hormone in growth hormone deficient children. Lancet 1:697-700.

Malozowski S, Tanner LA, Wysowski DK, Fleming GA, Stadel BV (1995) Benign intracranial hypertension in children with growth hormone deficiency treated with growth hormone. J Pediatr 126:996-999.

Maneatis T, Baptista J, Connelly K, Blethen S (2000) Growth hormone safety update from the National Cooperative Growth Study. J Pediatr Endocrinol Metab 13 Suppl 2:1035-1044.

Mills JL, Schonberger LB, Wysowski DK, Brown P, Durako SJ, Cox C, Kong F, Fradkin JE (2004) Long-term mortality in the United States cohort of pituitary-derived growth hormone recipients. J Pediatr 144:430-436.

Nishi Y, Tanaka T, Takano K, Fujieda K, Igarashi Y, Hanew K, Hirano T, Yokoya S, Tachibana K, Saito T, Watanabe S (1999) Recent status in the occurrence of leukemia in growth hormone-treated patients in Japan. GH Treatment Study Committee of the Foundation for Growth Science, Japan. J Clin Endocrinol Metab 84:1961-1965.

Sherman BM (1987) A National Cooperative Growth Study of Protropin. (Acta Paediatr Scand Suppl. 337:106-108.

Sklar CA, Mertens AC, Mitby P, Occhiogrosso G, Qin J, Heller G, Yasui Y, Robison LL (2002) Risk of disease recurrence and second neoplasms in survivors of childhood cancer treated with growth hormone: a report from the Childhood Cancer Survivor Study. J Clin Endocrinol Metab 87:3136-3141.

Swerdlow AJ, Higgins CD, Adlard P, Preece MA (2002) Risk of cancer in patients treated with human pituitary growth hormone in the UK, 1959-85: a cohort study. Lancet 360:273-277.

Taback SP, Dean HJ (1996) Mortality in Canadian children with growth hormone (GH) deficiency receiving GH therapy 1967-1992. The Canadian Growth Hormone Advisory Committee. J Clin Endocrinol Metab 81:1693-1696.

Taha DR, Bastian W, Castells S (2001) Growth hormone replacement therapy in children with leukemia in remission. Clin Pediatr (Phila) 40:441-445.

Wyatt D (2004) Lessons from the National Cooperative Growth Study. EurJ Endocrinol 151: S55-S59.

Appendix 1

Chronological List of NCGS Publications 1987-2004

Sherman BM. A National Cooperative Growth Study of Protropin. (1987) Acta Paediatr Scand Suppl 337:106-108.

August GP, Lippe BM, Blethen SL, Rosenfeld RG, Seelig SA, Johanson AJ, Compton PG, Frane JW, McClellan BH, Sherman BM. (1990) Growth hormone treatment in the United States: demographic and diagnostic features of 2331 children. J Pediatr 116:899-903.

Rosenfeld RG. Growth hormone therapy in Turner's syndrome: an update on final height. Genentech National Cooperative Study Group. (1992) Acta Paediatr Suppl Sep; 383:3-6.

Blethen SL, Compton P, Lippe BM, Rosenfeld RG, August GP, Johanson A. (1993) Factors predicting the response to growth hormone (GH) therapy in prepubertal children with GH deficiency. J Clin Endocrinol Metab 76:574-579.

Stabler B, Clopper RR, Siegel PT, Stoppani C, Compton PG, Underwood LE. (1994) Academic achievement and psychological adjustment in short children. The National Cooperative Growth Study. J Dev Behav Pediatr 15:1-6.

Carlsson LM, Attie KM, Compton PG, Vitangcol RV, Merimee TJ. (1994) Reduced concentration of serum growth hormone-binding protein in children with idiopathic short stature. National Cooperative Growth Study. J Clin Endocrinol Metab 78:1325-1330.

Kaplowitz PB. (1995) Effect of growth hormone therapy on final versus predicted height in short twelve- to sixteen-year-old boys without growth hormone deficiency. J Pediatr 126:478-480.

Vassilopoulou-Sellin, Klein MJ, Moore BD 3rd, Reid HL, Ater J, Zietz HA. Efficacy of growth hormone replacement therapy in children with organic growth hormone deficiency after cranial irradiation.(1995) Horm Res 43:188-193.

Tuffli GA, Johanson A, Rundle AC, Allen DB. Lack of increased risk for extracranial, nonleukemic neoplasms in recipients of recombinant deoxyribonucleic acid growth hormone. (1995) J Clin Endocrinol Metab 80:1416-1422.

Malozowski S, Tanner LA, Wysowski DK, Fleming GA, Stadel BV. Benign intracranial hypertension in children with growth hormone deficiency treated with growth hormone. (1995) J Pediatr 126:996-999.

Attie KM, Carlsson LM, Rundle AC, Sherman BM. Evidence for partial growth hormone insensitivity among patients with idiopathic short stature. The National Cooperative Growth Study. (1995) J Pediatr 127:244-250.

Hintz RL, Attie KM, Compton PG, Rosenfeld RG. (1995) Multifactorial studies of GH treatment of Turner syndrome: The Genentech National Cooperative Growth Study. In: Turner Syndrome in a Life-Span Perspective. Ed: K. Albertsson-Wikland and B. Lippe. Elsevier, Amsterdam pp 167-173.

Growth Hormone: Science, Research, and the NCGS. (1986) NCGS Advisory Group. Gardiner-Caldwell Synermed. Califon, New Jersey.

Blethen SL, Rundle AC. Slipped capital femoral epiphysis in children treated with growth hormone. A summary of the National Cooperative Growth Study experience. (1996) Horm Res 46:113-116.

Blethen SL, Allen DB, Graves D, August G, Moshang T, Rosenfeld R. (1996) Safety of recombinant deoxyribonucleic acid-derived growth hormone: The National Cooperative Growth Study experience. J Clin Endocrinol Metab 81:1704-1710.

Graves DA. Utility of the National Cooperative Growth Study database for safety reporting. (1996) J Pediatr. 128 (5 Pt 2):S1-3.

National Cooperative Growth Study: Ten Years of Guidance in Growth. Proceedings of a meeting. New York, October 12-15, 1995. (1996) J Pediatr 128 (5 Pt 2):S1-62.

Key LL Jr, Gross AJ. Response to growth hormone in children with chondrodysplasia.(1996) J Pediatr. 128 (5 Pt 2):S14-17.

Romano AA, Blethen SL, Dana K, Noto RA. Growth hormone treatment in Noonan syndrome: the National Cooperative Growth Study experience. (1996) J Pediatr 128 (5 Pt 2):S18-21.

Rogol AD, Breen TJ, Attie KM. National Cooperative Growth Study substudy. II: Do growth hormone levels from serial sampling add important diagnostic information? (1996) J Pediatr 128 (5 Pt 2):S42-46.

Chernausek SD, Breen TJ, Frank GR. Linear growth in response to growth hormone treatment in children with short stature associated with intrauterine growth retardation: the National Cooperative Growth Study experience. (1996) J Pediatr. 128 (5 Pt 2):S22-27.

Rotenstein D, Breen TJ. Growth hormone treatment of children with myelomeningocele. (1996) J Pediatr 128 (5 Pt 2):S28-31.

Blethen SL, Breen TJ, Attie KM. Overview of the National Cooperative Growth Study substudy of serial growth hormone measurements. (1996) J Pediatr 128 (5 Pt 2):S38-41.

Moshang T Jr, Rundle AC, Graves DA, Nickas J, Johanson A, Meadows A. Brain tumor recurrence in children treated with growth hormone: the National Cooperative Growth Study experience. (1996) J Pediatr. 128 (5 Pt 2):S4-7.

Rogol AD, Breen TJ, Attie KM. National Cooperative Growth Study substudy. II: Do growth hormone levels from serial sampling add important diagnostic information? (1996) J Pediatr 128 (5 Pt 2):S42-46.

Allen DB. Safety of human growth hormone therapy: current topics. (1996) J Pediatr 128 (5 Pt 2):S8-13.

Allen DB, Rundle AC, Graves DA, Blethen SL. Risk of leukemia in children treated with human growth hormone: review and reanalysis. (1997) J Pediatr 131 (1 Pt 2):S32-36.

Mentser M, Breen TJ, Sullivan EK, Fine RN. Growth-hormone treatment of renal transplant recipients: the National Cooperative Growth Study experience--a report of the National Cooperative Growth Study and the North American Pediatric Renal Transplant Cooperative Study. (1997) J Pediatr 131 (1 Pt 2):S20-24.

Allen DB, Rundle AC, Graves DA, Blethen SL. Risk of leukemia in children treated with human growth hormone: review and reanalysis. (1997) J Pediatr 131 (1 Pt 2):S32-36.

Attie KM, Julius JR, Stoppani C, Rundle AC. National Cooperative Growth Study substudy VI: the clinical utility of growth-hormone-binding protein, insulin-like growth factor I, and insulin-like growth factor-binding protein 3 measurements.(1997) J Pediatr 131 (1 Pt 2): S56-60.

Hardin DS, Sy JP. Effects of growth hormone treatment in children with cystic fibrosis: the National Cooperative Growth Study experience.(1997) J Pediatr 131 (1 Pt 2):S65-69.

Stephure DK, Blethen SL, McClellan BH. National Cooperative Growth Study substudy VIII: a new look at the natural history of short stature. (1997) J Pediatr 131 (1 Pt 2):S81-82.

Cronin MJ. Pioneering recombinant growth hormone manufacturing: pounds produced per mile of height. (1997) J Pediatr 131 (1 Pt 2):S5-7.

Goddard AD, Dowd P, Chernausek S, Geffner M, Gertner J, Hintz R, Hopwood N, Kaplan S, Plotnick L, Rogol A, Rosenfield R, Saenger P, Mauras N, Hershkopf R, Angulo M, Attie K. Partial growth-hormone insensitivity: the role of growth-hormone receptor mutations in idiopathic short stature. (1997) J Pediatr 131 (1 Pt 2):S51-55.

Proceedings of the National Cooperative Growth Study 10th Annual Investigators Meeting. San Francisco, California, October 17-20, 1996. Dedicated to the memory of Dr. Michael J. Cronin. (1997) J Pediatr 131 (1 Pt 2):S1-82.

Silverman BL, Friedlander JR. Is growth hormone good for the heart? (1997) J Pediatr 131 (1 Pt 2):S70-74.

Siegel PT, Clopper R, Stabler B. The psychological consequences of Turner syndrome and review of the National Cooperative Growth Study psychological substudy. (1998) Pediatrics 102 (2 Pt 3):488-491.

Stabler B, Siegel PT, Clopper RR, Stoppani CE, Compton PG, Underwood LE. Behavior change after growth hormone treatment of children with short stature. (1998) J Pediatr 133:366-373.

Root AW, Kemp SF, Rundle AC, Dana K, Attie KM. (1998) Effect of long-term recombinant growth hormone therapy in children--the National Cooperative Growth Study, USA, 1985-1994. J Pediatr Endocrinol Metab 11:403-412.

Plotnick L, Attie KM, Blethen SL, Sy JP. Growth hormone treatment of girls with Turner syndrome: the National Cooperative Growth Study experience. (1998) Pediatrics 102 (2 Pt 3):479-481.

Siegel PT, Clopper R, Stabler B. The psychological consequences of Turner syndrome and review of the National Cooperative Growth Study psychological substudy. (1998) Pediatrics 102 (2 Pt 3):488-491.

Rao JK, Julius JR, Breen TJ, Blethen SL. Response to growth hormone in attention deficit hyperactivity disorder: effects of methylphenidate and pemoline therapy. (1998) Pediatrics 102 (2 Pt 3):497-500.

August GP, Julius JR, Blethen SL. Adult height in children with growth hormone deficiency who are treated with biosynthetic growth hormone: the National Cooperative Growth Study experience. (1998) Pediatrics 102 (2 Pt 3):512-516.

MacGillivray MH, Blethen SL, Buchlis JG, Clopper RR, Sandberg DE, Conboy TA. Current dosing of growth hormone in children with growth hormone deficiency: how physiologic? (1998) Pediatrics 102 (2 Pt 3):527-530.

Bell JJ, Dana K. Lack of correlation between growth hormone provocative test results and subsequent growth rates during growth hormone therapy. (1998) Pediatrics 102 (2 Pt 3):518-520.

Diamond FB, Jorgensen EV, Root AW, Shulman DI, Sy JP, Blethen SL, Bercu BB. The role of serial sampling in the diagnosis of growth hormone deficiency. (1998) Pediatrics 102 (2 Pt 3):521-524.

Hintz RL. The role of auxologic and growth factor measurements in the diagnosis of growth hormone deficiency. (1998) Pediatrics 102 (2 Pt 3):524-526.

Allen DB, Julius JR, Breen TJ, Attie KM. (1998) Treatment of glucocorticoid-induced growth suppression with growth hormone. National Cooperative Growth Study. J Clin Endocrinol Metab 83:2824-2829.

National Cooperative Growth Study: Guidance in Growth. Proceedings of the National Cooperative 11th annual investigators meeting. Washington, DC, USA. September 25-28, 1997. (1998) Pediatrics 102 (2 Pt 3):iv, 479-530.

Kaplowitz PB, Rundle AC, Blethen SL. Weight relative to height before and during growth hormone therapy in prepubertal children. (1998) Horm Metab Res 30:565-569.

Root AW, Kemp SF, Rundle AC, Dana K, Attie KM. Effect of long-term recombinant growth hormone therapy in children – the National Cooperative Growth Study, USA, 1985-1994. (1998) J Pediatr Endocrinol Metab 11:403-412.

National Cooperative Growth Study: Guidance in Growth. Proceedings of the National Cooperative Growth Study 12th Annual Investigators Meeting. New Orleans, Louisiana, USA. October 8-11, 1998. (1999) Pediatrics 104 (4 Pt 2):999-1049.

Sandra L. Blethen. Leukemia in Children Treated with Growth Hormone. (1998) Trends in Endocrin Metabol 9:367-370.

Frindik JP, Kemp SF, Sy JP. Effects of recombinant human growth hormone on height and skeletal maturation in growth hormone-deficient children with and without severe pretreatment bone age delay. (1999) Horm Res 51:15-19.

Petryk A, Richton S, Sy JP, Blethen SL. The effect of growth hormone treatment on stature in Aarskog syndrome.(1999) J Pediatr Endocrinol Metab 12:161-165.

Castells S, Chakurkar A, Qazi Q, Bastian W. Robinow syndrome with growth hormone deficiency: treatment with growth hormone. (1999) J Pediatr Endocrinol Metab 12:565-571.

Frindik JP, Baptista J. (1999) Adult height in growth hormone deficiency: historical perspective and examples from the national cooperative growth study. Pediatrics 104 (4 Pt 2):1000-1004.

Kohn B, Julius JR, Blethen SL. (1999) Combined use of growth hormone and gonadotropin-releasing hormone analogues: the national cooperative growth study experience. Pediatrics 104 (4 Pt 2):1014-1018.

Bright GM, Julius JR, Lima J, Blethen SL. Growth hormone stimulation test results as predictors of recombinant human growth hormone treatment outcomes: preliminary analysis of the national cooperative growth study database. (1999) Pediatrics 104 (4 Pt 2):1028-1031.

Kemp SF, Sy JP. Analysis of bone age data from national cooperative growth study substudy VII. (1999) Pediatrics 104 (4 Pt 2):1031-1036.

Kaufman FR, Sy JP. Regular monitoring of bone age is useful in children treated with growth hormone. (1999) Pediatrics 104 (4 Pt 2):1039-1042.

Wyatt D. Melanocytic nevi in children treated with growth hormone. (1999) Pediatrics 104 (4 Pt 2):1045-1050.

Blackett PR, Rundle AC, Frane J, Blethen SL. Body mass index (BMI) in Turner Syndrome before and during growth hormone (GH) therapy. (2000) Int J Obes Relat Metab Disord 24:232-235.

August GP, Lightner ES, Root AW. Survey of clinical practice: growth hormone use during illness. (2000) J Pediatr Endocrinol Metab 13 Suppl 2:1031-1033.

Maneatis T, Baptista J, Connelly K, Blethen S. Growth hormone safety update from the National Cooperative Growth Study. (2000) J Pediatr Endocrinol Metab 13 Suppl 2:1035-1044.

Wright NM. Just taller or more bone? The impact of growth hormone on osteogenesis imperfecta and idiopathic juvenile osteoporosis.(2000) J Pediatr Endocrinol Metab 13 Suppl 2:999-1002.

Reiter EO, Blethen SL, Baptista J, Price L. Early initiation of growth hormone treatment allows age-appropriate estrogen use in Turner's syndrome. (2001) J Clin Endocrinol Metab 86:1936-1941.

Taha DR, Bastian W, Castells S. Growth hormone replacement therapy in children with leukemia in remission. (2001) Clin Pediatr (Phila) 40:441-445.

Kemp SF, Alter CA, Dana K, Baptista J, Blethen SL. Use of magnetic resonance imaging in short stature: data from National Cooperative Growth Study (NCGS) Substudy 8. (2002) J Pediatr Endocrinol Metab 15 Suppl 2:675-679.

Meacham LR, Sullivan K. Characteristics of growth hormone therapy for pediatric patients with brain tumors in the National Cooperative Growth Study (NCGS) and from a survey of pediatric endocrinologists. (2002) J Pediatr Endocrinol Metab 15 Suppl 2:689-696.

Parker KL, Hunold JJ, Blethen SL. Septo-optic dysplasia/optic nerve hypoplasia: data from the National Cooperative Growth Study (NCGS). (2002) J Pediatr Endocrinol Metab 15 Suppl 2:697-700.

Reeves GD, Doyle DA. Growth hormone treatment and pseudotumor cerebri: coincidence or close relationship? (2002) J Pediatr Endocrinol Metab 15 Suppl 2:723-730.

Frindik JP, Kemp SF, Hunold JJ. Near adult heights after growth hormone treatment in patients with idiopathic short stature or idiopathic growth hormone deficiency. (2003) J Pediatr Endocrinol Metab 16 Suppl 3:607-612.

Levy RA, Connelly K. Diagnostic growth hormone deficiency testing practices among patients in the NCGS/NCSS databases. (2003) J Pediatr Endocrinol Metab 16 Suppl 3:619-624.

Parker KL, Wyatt DT, Blethen SL, Baptista J, Price L. Screening girls with Turner syndrome: the National Cooperative Growth Study experience. (2003) J Pediatr 143:133-135.

Proceedings of the 16th Annual National Cooperative Growth Study (NCGS) and the 3rd Annual National Cooperative Somatropin Surveillance (NCSS) Investigator Meeting. Chicago, Illinois, USA. October 17-20, 2002. (2003) J Pediatr Endocrinol Metab 16 Suppl 3:585-690.

Bell JJ, August GP, Blethen SL, Baptista J. Neonatal hypoglycemia in a growth hormone registry: incidence and pathogenesis. (2004) J Pediatr Endocrinol Metab 17:629-635.

Wyatt D. Lessons from the national cooperative growth study. (2004) Eur J Endocrinol 151 Suppl 1:S55-59.

Why we are Treating Children with Growth Hormone: Lessons from the French Registry

Jean Claude Carel and Jean-Louis Chaussain[1]

Summary

Recombinant growth hormone has been used for approximately two decades and is now widely used worldwide in a variety of situations. Post-marketing data collected by growth hormone manufacturers have provided an enormous amount of information on the safety and efficacy of these treatments. In addition to this approach, national registries have provided data on a population basis. In France, the France-Hypophyse Association has collected data up to 1997 and has produced several analyses, which are summarized here.

Introduction

Recombinant growth hormone (GH) has been used for approximately two decades and is now widely used worldwide in a variety of situations in children. Altogether, the indications for its use have evolved from treating severe GH deficiency, scarce extractive GH, to non-GH deficient indications. Its use has expanded from specific conditions such as Turner syndrome or chronic renal failure to wider ones, such as treating short children born small for gestational age or with idiopathic short stature. Although these applications have been supported by clinical trials that have led to administrative approval, the amount of data available at the time of approval was often limited. In particular, adult height data were not available when the earlier indications were obtained in the mid 1980s and early 1990s. This situation highlights the need for additional post-marketing data to evaluate the long-term effects on growth and on additional clinical endpoints and to assess long-term safety.

Post-marketing databases such as the KIGS and the NCGS have been developed and have provided extensive information on several aspects of GH use. Salient results from the NCGS database are presented in this volume by Dr. Hintz. Although these databases effectively monitor efficacy and safety while patients are undergoing treatment, one serious limitation is the high number of losses of follow-up of patients who therefore fail to attain the outcomes. In particular,

[1] Department of Pediatric Endocrinology and INSERM U561, Groupe hospitalier Cochin-Saint Vincent de Paul and Faculté Cochin - Université Paris V, Paris, France

Carel et al.
Deciphering Growth
© Springer-Verlag Berlin Heidelberg

adult height analyses tend to concentrate on patients who are followed extensively, whereas patients who interrupt the treatment early are not taken into account. In this context, national databases can provide additional insights. They are less limited by regulatory concerns and can follow patients after they have ceased treatment. From 1973 to 1997, the prescription of GH in France had to be individually approved by a central agency (Association France-Hypophyse). This restriction has facilitated the analysis of population-based cohort of patients treated for GH deficiency or Turner syndrome.

Growth hormone deficiency

Idiopathic GH deficiency is the indication for treatment in 50% of the children receiving GH, as reported for 100,000 children worldwide in 1999 (Guyda 1999). GH treatments aim to normalize height, correct health problems associated with GH deficiency and achieve an adult height in the normal range for the general population and for familial genetic potential (The drug and therapeutic committee of the Lawson Wilkins pediatric endocrine society 1995; Brook 1997; Vance and Mauras 1999). Although GH has been used for four decades, initially as an extract and now in recombinant form, its long-term effects on adult height are still not fully appreciated (Guyda 1999). No long-term controlled trial has been performed and evaluation is based on comparisons with historical controls or on changes in height (Cole 1993; Brook et al. 2000). GH deficiency is poorly defined and ranges from severe to borderline. The issue of diagnostic criteria for GH deficiency has been widely addressed (Rosenfeld et al. 1995; Carel et al. 1997; August 1998; Saggese et al. 1998; Guyda 2000), but profiles of patients treated around the world do not correspond to strict definitions, with little change over time. Long-term follow-up is required to analyze adult height. Adult heights are generally recorded for patients who have been followed extensively, excluding those who stop treatment prematurely, therefore giving biased results (Coste et al. 1997).

Most published studies in GH deficiency have reported short-term (one to two years) effects of GH. Most long-term results published in the 1990s concerned small groups of patients (reviewed in Guyda 1999). Co-operative studies have reported results for a small proportion (<5%) of the patients enrolled and analyses are therefore prone to selection bias (Blethen et al. 1997; August et al. 1998; Cutfield et al. 1999).

In 1999, we collected adult height data from a population-based cohort of patients with a diagnosis of idiopathic GH deficiency who had started their treatment between 01/07/1987 and 31/12/1992, and who had attained their adult height by September 1999. Details of the study were published in 2002 (Carel et al. 2002); we will briefly summarize them here. The mean age at onset of treatment (12.6 years) and the fact that more than 90% of the children had stimulated GH peaks over 5 ng/ml classify them more accurately in the idiopathic short stature group than in the true GH-deficient group (Marin et al. 1994; Rosenfeld et al. 1995). As shown in Figure 1, we classified patients into those who had completed their treatment until the near-end of growth (roughly 50% of the

2852 patients followed to adult height) and those who had stopped treatment at various time points before reaching this stage. In the direct analysis of data, all groups gained about 1.1 SDS, raising the question whether this height gain was due to spontaneous catch-up in individuals with delayed puberty or to the effect of growth hormone. In particular, the patients who had used GH for the shortest period of time (less than 18 months) experienced similar results as those who used treatment for the longest periods. Using multivariate analysis, we tried to take into account several factors associated with growth, resulting in a model explaining 58% of the variance of height gain expressed in SDS. Most of the factors identified were unrelated to the treatment and only 4% of the variance was explained by treatment variables. Quite interestingly, completion and duration of the treatment had opposite effects, with children with "incomplete" treatments gaining more height and with longer treatments associated with higher gains (Carel et al. 2002). The mean effect was close to 1 cm of adult height gain per year of treatment. Other than the limits of a multicenter observational study, the two caveats of our study are first, the relative heterogeneity of the patients who were mostly selected by their height and their (unreliable) response to GH provocative stimuli (Carel et al. 1997) and second, the relatively low dose of GH used (0.4 U/kg/wk or 0.02 mg/kg/wk).

It should also be kept in mind that the design of the study concentrated on the older patients at onset of GH treatment and therefore excluded those with early onset GH deficiency. In addition, patients with organic GH deficiency were not analyzed and are undoubtedly those who benefit most from GH treatment.

Turner syndrome

Turner syndrome, first described in 1938 (Turner 1938), is a common chromosomal disorder, affecting approximately 1 in every 2,500 liveborn females. It results from the partial or total absence of one of the X chromosomes (Sybert and McCauley 2004). Short stature is a common feature of Turner syndrome; adult patients have a mean height approximately 20 cm less than that for unaffected women of the same ethnic group (Lyon et al. 1985; Cabrol et al. 1996; Sybert and McCauley 2004). Short stature results partly from haploinsufficiency of the SHOX gene on the distal part of the short arm of chromosome X; the GH/IGF-I axis is normal in Turner syndrome (Sybert and McCauley 2004).

Treatment with recombinant human GH has been offered to most affected children since the early 1990s and is now considered standard (Ranke and Saenger 2001). This treatment has been shown to increase growth rate in the short term (Rosenfeld et al. 1992). Comparisons of adult heights with pre-treatment predicted heights based on disease-specific normative data (Lyon et al. 1985; Sempé et al. 1996) have shown variable outcomes ranging from no effect (Taback et al. 1996; Chu et al. 1997; Dacou-Voutetakis et al. 1998) to a mean increase of up to 16.9 cm (Carel et al. 1998; Plotnick et al. 1998; Rosenfeld et al. 1998; Chernausek et al. 2000; Reiter et al. 2001; Quigley et al. 2002; Ranke et al. 2002; Massa et al. 2003; van Pareren et al. 2003). This variability may be accounted for by several factors, including age at GH initiation, ethnic origin, GH

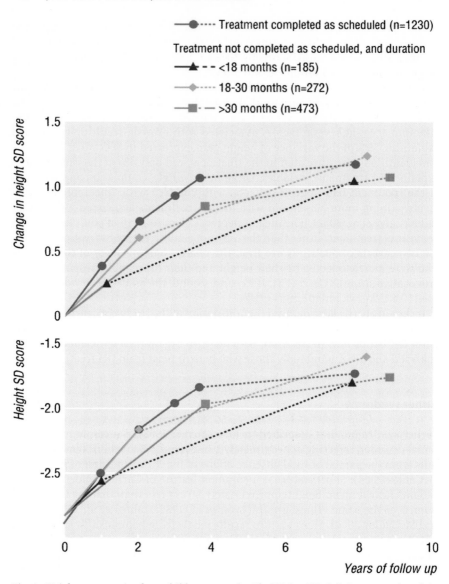

Fig. 1. Height outcome in short children treated with GH for GH deficiency: results of the France-Hypophyse database (adapted from Carel et al. 2002).

dose and pubertal management. Most adult height reports have been based on clinical trials for which generalization may not be valid (Carel et al. 1998; Rosenfeld et al. 1998; Chernausek et al. 2000; Quigley et al. 2002; van Pareren et al. 2003), whereas others have been based on large post-marketing databases with potential biases (particularly completion bias, where those who stay in the study until adult height might have a better outcome than those who stop treatment and are lost to follow-up (Plotnick et al. 1998; Ranke et al. 2002). More recently, a Canadian randomized trial has confirmed that GH treatment increased adult height in Turner syndrome (Stephure 2005).

Gonadal dysgenesis is another key feature of Turner syndrome. Spontaneous pubertal development occurs in only about 20% of patients, with 2-5% experiencing spontaneous menarche (Pasquino et al. 1997); the vast majority of patients with Turner syndrome require treatment with estrogens and progestin to achieve adequate pubertal development (Sybert and McCauley 2004). Estrogens have been shown to be involved in epiphyseal fusion (Grumbach and Auchus 1999). This finding has called into question the timing and means of sex-steroid treatment in adolescents with Turner syndrome, with some experts advocating early pubertal induction and others late pubertal induction, based on psychosocial or auxological issues (Ross et al. 1996; Chernausek et al. 2000; Hogler et al. 2004).

Adult height in Turner syndrome

The France-Hypophyse database also allowed us to evaluate adult height and health-related quality of life in patients with Turner syndrome. Adult height data were available for 704 of the 891 eligible patients (79%; Soriano-Guillen et al. 2005). GH was used at the dose of 0.8 ± 0.2 IU/kg/wk (0.26 ± 0.06 mg/kg/wk) (M ± SD) for 5.0 ± 2.2 years. Puberty was classified as spontaneous (10%), spontaneous with secondary estrogens (13%) or induced (77%). Estrogen treatment was initiated at 15.0 ± 1.9 years of age in those with induced puberty. Mean adult height was 149.9 ± 6.1 cm, 8.5 cm above projected height. The model explained 90% of variance, with major effects of age at initiation and duration of treatment. Other factors included birth length, target height, bone age delay and weight at initiation of treatment, age at pubertal onset, GH dose and number of injections per week. Age at introduction of estrogens was not a predictor and the use of percutaneous vs oral estrogens was associated with greater height (+2.1 cm, 95% CI 1.00 to 3.25). Our results therefore supported the early initiation of GH treatment and induction of puberty at a physiological age to achieve optimal adult height. They suggested that GH should be injected daily and percutaneous estrogens should be used, an observation that had not been made in previous analyses.

Quality of life in Turner syndrome

We also analyzed quality of life using standardized questionnaires such as the SF-36 and GHQ-12 (Carel et al. 2005). Treatments to promote growth in girls with Turner's syndrome aim to reduce the impact of short stature on psychosocial functioning and quality of life. However, these aspects have not been evaluated in young adults after treatment completion. Several small studies have reported changes in psychosocial functioning and quality of life in adolescents and women with Turner's syndrome (reviewed in Boman et al. 1998; Elsheikh et al. 2002). However, the effects on psychosocial functioning of height or height gain from GH treatment remain unclear (Lagrou et al. 1998; Siegel et al. 1998). SF-36 scores in the 568 women with Turner syndrome were similar to those of women of the same age from the general population (Fig. 2). If expressed as SDS,

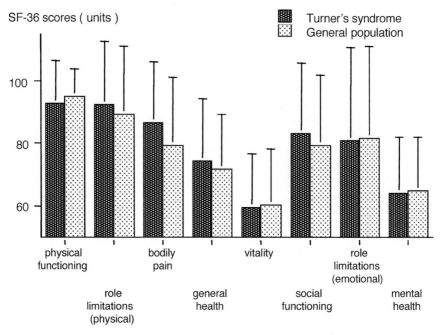

Fig. 2. Quality of life measurements in Turner's syndrome patients. Results are shown as absolute scores for patients with Turner's syndrome (▨) or for French women from the general population aged 18-24 years (▨); means ± SDs are shown (adapted from Carel et al. 2005).

these scores did not differ significantly from zero. The proportion of women with Turner's syndrome who had a high GHQ-12 score (≥3) was lower than that for the general population (24% vs 31%, respectively, p<0.001). We assessed the effect of several of the patients' characteristics on SF-36 scores. The father's socio-professional background (manual vs non manual) and the patient's level of education (secondary education completed or not) were associated with several quality of life dimensions. GHQ-12 scores were correlated with SF-36 scores. Further analyses were performed after adjustment for these three potentially confounding variables (paternal socio-professional status, participant's educational level and GHQ-12 score). Adult height was not associated with quality of life, regardless of whether height was treated as a continuous variable or broken down into categories. Similarly, duration of treatment, age at treatment initiation, total GH dose and estimated height gain were not associated with quality of life. The patients who had the highest expectations regarding GH treatment had the lowest quality of life scores. Cardiac involvement was strongly associated with low physical scores, whereas kidney and genital malformations associated with the presence of Y chromosome material were not. Unexpectedly, otological involvement, present in 26% of the patients and either detected during childhood care or declared by the patients at the time of the survey, was strongly associated with perceived health-related quality of life in all but one dimension. Otological conditions resulted in a loss of 4 to 13 score points or 0.2 to 0.7 SDS units. Patients whose puberty had been induced after the age of 15 years had significantly lower general health perception scores. Other factors were analyzed and found not to be associated with quality of life as assessed by the SF-36 questionnaire: karyotype, dysmorphic features of Turner's syndrome, sexual intercourse experience, presence of thyroid dysfunction or self-reported visual problems. We concluded from these studies that quality of life is normal and unaffected by height in young adults with Turner syndrome treated with GH.

These data emphasize the need to give appropriate attention to general health and otological care rather than focus on stature in the care of children with Turner's syndrome.

Conclusion

Although the France-Hypophyse Association interrupted its activities in 1997, valuable data have been collected through this database. In the future, the availability of this database allows the possibility of collecting long-term data on tolerance to recombinant GH.

Acknowledgments

The studies were supported by grants from Ministère de la Santé, Programme Hospitalier de Recherche Clinique AOM96016 and AOM98009.

Dr Joel Coste and Emmanuel Ecosse are acknowledged for their role in the design of the studies and statistical analysis. We would like to thank Dr Irène Bastie-Sigeac, Vean Eng Ly and Sabine Ximenes for their invaluable contributions. We also thank the coinvestigators, Dr Raja Brauner, Sylvie Cabrol, Pierre Chatelain, Paul Czernichow, Juliane Léger, Marc Nicolino, Maité Tauber and all the physicians involved in the follow-up of patients and in the review process at Association France-Hypophyse.

References

August GP (1998) Symposium on controversies in the diagnosis of growth hormone deficiency. Pediatrics 102 (2 Pt 3): 517.

August GP, Julius JR, Blethen SL (1998) Adult height in children with growth hormone deficiency who are treated with biosynthetic growth hormone: the National Cooperative Growth Study experience. Pediatrics 102 (2 Pt 3): 512-516.

Blethen SL, Baptista J, Kuntze J, Foley T, LaFranchi S, Johanson A, (1997) Adult height in growth hormone (GH)-deficient children treated with biosynthetic GH. The Genentech Growth Study Group. J Clin Endocrinol Metab 82: 418-420.

Boman UW, Moller A, Albertsson-Wikland K (1998) Psychological aspects of Turner syndrome. J Psychosom Obstet Gynaecol 19: 1-18.

Brook CGD (1997) Growth hormone: panacea or punishment for short stature? Brit Med J 315: 692-693.

Brook CGD, Kelnar CJH, Betts P (2000) Controversy: which children should receive growth hormone treatment. Arch Dis Child 83: 176-178.

Cabrol S, Saab C, Gourmelen M, Raux-Demay MC, Le Bouc Y (1996) Syndrome de Turner: croissance staturopondérale et maturation osseuses spontanées. Arch Pediatr 3: 313-318.

Carel JC, Mathivon L, Gendrel C, Ducret JP, Chaussain JL (1998) Near normalisation of final height with adapted doses of growth hormone in Turner's syndrome. J Clin Endocrinol Metab 83: 1462-1466.

Carel JC, Tresca J-P, Letrait M, Le Bouc Y, Job J-C, Chaussain JL, Coste J (1997) Growth hormone testing for the diagnosis of growth hormone deficiency in childhood: a population register-based study. J Clin Endocrinol Metab 82: 2117-2121.

Carel JC, Ecosse E, Nicolino M, Tauber M, Leger J, Cabrol S, Bastié-Sigeac I, Chaussain JL, Coste J (2002) Adult height after long-term recombinant growth hormone treatment for idiopathic isolated growth hormone deficiency: observational follow-up study of the French population-based registry. Brit Med J 325: 70-73.

Carel JC, Ecosse E, Bastie-Sigeac I, Cabrol S, Tauber M, Leger J, Nicolino M, Brauner R, Chaussain JL, Coste J (2005) Quality of life determinants in young women with Turner's syndrome after growth hormone treatment: results of the StaTur population-based cohort study. J Clin Endocrinol Metab 90: 1992-1997.

Chernausek SD, Attie KM, Cara JF, Rosenfeld RG, Frane J (2000) Growth hormone therapy of Turner syndrome: the impact of age of estrogen replacement on final height. Genentech, Inc., Collaborative Study Group. J Clin Endocrinol Metab 85: 2439-2445.

Chu CE, Paterson WF, Kelnar CJ, Smail PJ, Greene SA, Donaldson MD (1997) Variable effect of growth hormone on growth and final adult height in Scottish patients with Turner's syndrome. Acta Paediatr 86: 160-164.

Cole T (1993) Methodology for the analysis of longitudinal height data during puberty. KIGS Report 10: 37-44.

Coste J, Letrait M, Carel JC, Tresca J-P, Chatelain P, Rochiccioli P, Chaussain JL, Job J-C (1997) Long term results of GH treatment in short stature children: a population register-based study. Brit Med J 315: 708-713.

Cutfield W, Lindberg A, Albertsson Wikland K, Chatelain P, Ranke MB, Wilton P (1999) Final height in idiopathic growth hormone deficiency: the KIGS experience. KIGS International Board. Acta Paediatr Suppl 88: 72-75.

Dacou-Voutetakis C, Karavanaki-Karanassiou K, Petrou V, Georgopoulos N, Maniati-Christidi M, Mavrou A (1998) The growth pattern and final height of girls with Turner syndrome with and without human growth hormone treatment. Pediatrics 101: 663-668.

Elsheikh M, Dunger DB, Conway GS, Wass JA (2002) Turner's syndrome in adulthood. Endocr Rev 23: 120-140.

Grumbach MM, Auchus RJ (1999) Estrogen: consequences and implications of human mutations in synthesis and action. J Clin Endocrinol Metab 84: 4677-4694.

Guyda HJ (1999) Commentary. Four decades of growth hormone therapy for short children: what have we achieved? J Clin Endocrinol Metab 84: 4307-4316.

Guyda HJ (2000) Commentary. Growth hormone testing and the short child. Pediatr Res 48: 579-580.

Hogler W, Briody J, Moore B, Garnett S, Lu PW, Cowell CT (2004) Importance of estrogen on bone health in Turner syndrome: a cross-sectional and longitudinal study using dual-energy X-ray absorptiometry. J Clin Endocrinol Metab 89: 193-199.

Lagrou K, Xhrouet-Heinrichs D, Heinrichs C, Craen M, Chanoine JP, Malvaux P, Bourguignon JP (1998) Age-related perception of stature, acceptance of therapy, and psychosocial functioning in human growth hormone-treated girls with Turner's syndrome. J Clin Endocrinol Metab 83: 1494-1501.

Lyon AJ, Preece MA, Grant DB (1985) Growth curve for girls with Turner syndrome. Arch Dis Child 60: 932-935.

Marin G, Domene H, Barnes K, Blackwell B, Cassorla F, Cutler G. Jr. (1994) The effects of estrogen priming and puberty on the growth hormone response to standardized treadmill exercise and arginine-insulin in normal girls and boys. J Clin Endocrinol Metab 79: 537-541.

Massa G, Heinrichs C, Verlinde S, Thomas M, Bourguignon JP, Craen M, Francois I, Du Caju M, Maes M, De Schepper J (2003) Late or delayed induced or spontaneous puberty in girls with Turner syndrome treated with growth hormone does not affect final height. J Clin Endocrinol Metab 88: 4168-4174.

Pasquino AM, Passeri F, Pucarelli I, Segni M, Municchi G (1997) Spontaneous pubertal development in Turner's syndrome. Italian Study Group for Turner's Syndrome. J Clin Endocrinol Metab 82: 1810-1813.

Plotnick L, Attie KM, Blethen SL, Sy JP (1998) Growth hormone treatment of girls with Turner syndrome: the National Cooperative Growth Study experience. Pediatrics 102 (2 Pt 3): 479-481.

Quigley CA, Crowe BJ, Anglin DG, Chipman JJ (2002) Growth hormone and low dose estrogen in Turner syndrome: results of a United States multi-center trial to near-final height. J Clin Endocrinol Metab 87: 2033-2041.

Ranke MB, Saenger P (2001) Turner's syndrome. Lancet 358: 309-314.

Ranke MB, Partsch CJ, Lindberg A, Dorr HG, Bettendorf M, Hauffa BP, Schwarz HP, Mehls O,m Sander S, Stahnke N, Steinkamp H, Said E, Sippell W (2002) Adult height after GH therapy in 188 Ullrich-Turner syndrome patients: results of the German IGLU Follow-up Study 2001. Eur J Endocrinol 147: 625-633.

Reiter EO, Blethen SL, Baptista J, Price L (2001) Early initiation of growth hormone treatment allows age-appropriate estrogen use in Turner's syndrome. J Clin Endocrinol Metab 86: 1936-1941.

Rosenfeld RG, Frane J, Attie KM, Brasel JA, Burstein S, Cara JF, Chernausek S, Gotlin RW, Kuntze J, Lippe BM, Mahoney PC, Moore WV, Saenger P, Johanson AJ (1992) Six-year results of a randomized, prospective trial of human growth hormone and oxandrolone in Turner syndrome. J Pediatr 121: 49-55.

Rosenfeld RG, Attie KM, Frane J, Brasel JA, Burstein S, Cara JF, Chernausek S, Gotlin RW, Kuntze J, Lippe BM, Mahoney CP, Moore WV, Saenger, P, Johanson AJ (1998) Growth hormone therapy of Turner's syndrome: beneficial effect on adult height. J Pediatr 132: 319-324.

Rosenfeld RG, Albertsson-Wikland K, Cassorla F, Frasier SD, Hasegawa Y,Hintz RL, LaFranchi S, Lippe BM, Loriaux DL, Melmed S, Preese MA, Ranke MB, Reiter EO, Rogol AD, Underwood LE, Werther GE (1995) Diagnostic controversy: the diagnosis of childhood growth hormone deficiency revisited. J Clin Endocrinol Metab 80: 1532-1540.

Ross JL, McCauley E, Roeltgen D, Long L, Kushner H, Feuillan P, Cutler GB, Jr. (1996) Self-concept and behavior in adolescent girls with Turner syndrome: potential estrogen effects. J Clin Endocrinol Metab 81: 926-931.

Saggese G, Ranke MB, Saenger P, Rosenfeld RG, Tanaka T, Chaussain JL, Savage MO (1998) Diagnosis and treatment of growth hormone deficiency in children and adolescents: towards a consensus. Ten years after the Availability of Recombinant Human Growth Hormone Workshop, Pisa, Italy, 27-28 March 1998. Horm Res 50: 320-340.

Sempé M, Hansson Bondallaz C, Limoni C (1996) Growth curves in untreated Ullrich-Turner syndrome: French reference standards 1-22 years. Eur J Pediatr 155: 862-869.

Siegel PT, Clopper R, Stabler B (1998) The psychological consequences of Turner syndrome and review of the National Cooperative Growth Study psychological substudy. Pediatrics 102(2 Pt 3): 488-491.

Soriano-Guillen L, Coste J, Ecosse E, Leger J, Tauber M, Cabrol S, Nicolino M, Brauner R The StaTur study group, Chaussain JL, Carel JC (2005) Adult height and pubertal growth in Turner's syndrome after treatment with recombinant growth hormone: observational follow-up study of the StaTur population-based cohort. J Clin Endocrinol Metab, in press.

Stephure DK (2005) Impact of growth hormone supplementation on adult height in turner syndrome: results of the Canadian randomized controlled trial. J Clin Endocrinol Metab 90: 3360-3366.

Sybert VP, McCauley E (2004) Turner's syndrome. New Engl J Med 351: 1227-1238.

Taback SP, Collu R, Deal CL, Guyda HJ, Salisbury S, Dean HJ, Van Vliet G (1996) Does growth-hormone supplementation affect adult height in Turner's syndrome? Lancet 348: 25-27.

The drug and therapeutic committee of the Lawson Wilkins pediatric endocrine society (1995) Guidelines for the use of growth hormone in children with short stature. J Pediatr 127: 857-867.

Turner HH (1938) A syndrome of infantilism, congenital webbed neck and cubitus valgus. Endocrinology 23: 566.

van Pareren YK, de Muinck Keizer-Schrama SM, Stijnen T, Sas TC, Jansen M, Otten BJ, Hoor-weg-Nijman JJ, Vulsma T, Stokvis-Brantsma WH, Rouwe CW, Reeser HM, Gerver WJ, Gos-en JJ, Rongen-Westerlaken C, Drop SL (2003) Final height in girls with Turner syndrome after long-term growth hormone treatment in three dosages and low dose estrogens. J Clin Endocrinol Metab 88: 1119-1125.

Vance ML, Mauras N (1999) Drug therapy: Growth hormone therapy in adults and children. New Engl J Med 341: 1206-1216.

Subject Index

Printing: Krips bv, Meppel
Binding: Stürtz, Würzburg